Computer-aided Analysis and Design of Electronic Networks

Computer-aided Analysis and Design of Electronic Networks

J. B. Grimbleby
University of Reading

PITMAN PUBLISHING
128 Long Acre, London WC2E 9AN
A Division of Longman Group UK Limited

© J. B. Grimbleby 1990

First published in Great Britain 1990

British Library Cataloguing in Publication Data

Grimbleby, J. B.
  Computer-aided analysis and design of electronic networks.
  1. Electronic equipment. Circuits. Design. Application of
  computer systems
  I. Title
  621.3815

ISBN 0-273-03148-1

ISBN 0 273 03148 1

Contents

Preface

This is a book about the use of computers to analyze and design electronic networks. It is concerned with both the theoretical foundations and the practical computational aspects of a number of analysis and design techniques. Although it is primarily intended to be an undergraduate text, some of the material should be of interest to postgraduates and also to practising electronic engineers.

There are often several different ways of performing a particular type of network analysis. For example, the gain and phase of a network at some specified frequency can be determined by the use of either numerical or symbolic network analysis. The choice of the most suitable method often rests on the accuracy and efficiency of the numerical techniques employed in the analysis. Consequently it has been the aim throughout this book to integrate the network theory and the numerical aspects of computer-aided network analysis and design.

Nowadays electronic engineering calculations are usually performed on personal computers or workstations, many of which are as powerful as the mainframe computers of only a few years ago. Personal computers offer a high degree of interactivity and fast graphical displays at a reasonable cost. IBM introduced their personal computer in 1981 and this rapidly became a *de facto* standard; nearly all personal computers are now compatible with the IBM-PC. This standardization acted as a catalyst to software manufacturers who soon produced a range of affordable engineering software to run on personal computers. In particular, compilers became available for all the major computing languages.

Algorithms developed in this book are illustrated by sections of programs written in the computer language Modula-2. The choice of a suitable computer language was not an easy one: it had to be a modern language encouraging structured programming, easy to use, and available on personal computers at reasonable cost. Pascal, Modula-2 and Ada were all considered as possible candidates. Pascal is widely available and is suitable as an initial teaching language, but lacks modularity. On the other hand, Ada is extremely powerful, but validated Ada compilers for personal computers tend to be expensive and less efficient than compilers for other languages. These considerations led to the selection of Modula-2. Fortunately all these languages are quite similar and anyone familiar with Pascal or Ada will have no difficulty in understanding the Modula-2 programs presented here.

Most of this book is concerned with applications of computers to network analysis. Methods for calculating the frequency and time-domain responses of linear networks, and the d.c. behaviour of non-linear networks, are described. Techniques for determining the effects of component tolerances are also discussed. With the exception of symbolic analysis, all of the techniques presented are based on numerical procedures. Computer analysis is used indirectly in the design process, allowing the performance of different designs to be assessed. By contrast, computer optimization involves computers directly in the design process. Methods for optimization of single and multi-variable systems are described. Finally, no book on computer-aided analysis and design would be complete without a reference to SPICE, and an appendix is devoted to the use of this widely available network analysis program.

Prerequisites to studying this book are a knowledge of basic network theory, including network theorems and the Laplace transform, and a familiarity with elementary matrix theory, differential equations and statistics.

I should like to express my gratitude to Professor Eric Faulkner for the many hours spent discussing (and arguing about) matters related to computer-aided design. My thanks also go to my family for their encouragement and support during the preparation of this book, and especially to my wife Linda who word-processed the manuscript and assisted in the proof-reading.

James B. Grimbleby
Reading, 1990

The Modula-2 source code of the programs described in this book can be obtained on IBM-PC compatible floppy discs from:

Dr. J. B. Grimbleby,
Department of Engineering,
University of Reading,
Whiteknights,
Reading RG6 2AY.

Please enclose payment of £10 to cover administrative costs, and state clearly the preferred disc format ($5\frac{1}{4}$ inch 360 kbyte or $3\frac{1}{2}$ inch 1.4 Mbyte).

1 Introduction

1.1 The PC Revolution

Before about 1960 the only available methods for performing engineering calculations were slow and laborious, involving logarithm tables, slide rules or mechanical calculators. Then in the early 1960s the first computer revolution took place. Computers, which until this time had been confined to research laboratories, become more widely available. Universities and large commercial organizations installed computers in specially designed air-conditioned computer centres.

Computer users were required to submit their programs on punched paper tape or punched cards, which were then placed in a queue and run, one at a time, on the computer. Some hours or days later the results would become available. Often a program would fail to run because of a simple syntax error, such as a missing comma, and after correction would have to be re-submitted. This meant that the time taken to develop even a relatively simple program could run into days or weeks. Computers were normally programmed in assembly code or Fortran, which was the only widely available high-level language at the time. In spite of this, the use of computers increased rapidly and they soon became an indispensable tool for performing engineering calculations.

The next big advance was the advent of multi-user operating systems in the late 1960s. These made it possible for a computer user to sit at a typewriter-like console connected to the computer, and to edit and run programs concurrently with other users. A special language, Basic, was developed for use in this type of environment, although other languages such as Fortran and Pascal are also suitable. Working with a computer interactively resulted in an enormous improvement in productivity; programs which had previously taken days or weeks to develop could now be produced in a matter of hours.

Consoles based on mechanical printers were not particularly convenient for inputting and editing programs, and during the 1970s they were gradually replaced by visual display units (VDUs) using raster-scan cathode-ray tubes. These allowed much faster communication rates between console

1

and computer, and made possible the use of screen editors and interactive graphics.

The PC revolution began in 1981. In that year IBM, the world's largest manufacturer of mainframe computers, introduced the IBM personal computer, or PC. It is true that some microprocessor-based computers such as the Apple had been available before the IBM-PC, but these were generally small machines designed to be programmed in Basic. Based on the 8088 microprocessor (a device whose accumulator architecture is derived from one of the earliest microprocessors, the 8080) and initially with only a limited amount of memory, the IBM-PC was nevertheless an immediate success. Before long other manufacturers began to produce computers that were compatible with the IBM-PC. These computers were able to run software, and to accept hardware such as plug-in interface cards, designed for the IBM-PC. The IBM-PC soon became the *de facto* standard for personal computers.

Software manufacturers, encouraged by this standardization, started to devote considerable resources to developing software for what was obviously going to be a huge market, and a wide range of software for commercial, scientific and engineering applications soon became available. In particular compilers were produced for all the major computer languages, including Basic, Fortran, C, Pascal and Modula-2.

It is a well-established rule of computing that the amount of computation rapidly expands to fill the available resources, and this is as true of personal computers as it is of mainframe computers. Users of personal computers began to demand more computational power. Faster microprocessors, using higher clock speeds and 16-bit or 32-bit busses, have gradually replaced the original 4·77 MHz 8-bit bus 8088 processor. Winchester technology hard discs offer faster access and much greater capacity than floppy discs. Finally the amount of random-access memory has increased to the maximum of 640 kbyte imposed by the limited addressing capability of the 8088 and compatible microprocessors.

As a result of these developments most electronic computer-aided analysis and design can now be performed on local personal computers, rather than on remote mainframes. Distributing computing power amongst the users, rather than centralizing it in mainframe computers, has many advantages. More computing power is generally available from a number of personal computers than from a single mainframe of similar cost. Personal computers also tend to be fairly reliable; if a fault does develop it affects only the user of a particular machine. Perhaps the most important advantage is the high level of interaction that is available on a personal computer. Close coupling of processor and display allows complex graphical displays to be quickly updated and a pointing device, such as a mouse, to be used for rapid positioning of objects on the screen, or the selection of items from a menu.

1.2 Design Using Computers

Computers are involved in many aspects of everyday life and come in all shapes and sizes. At one extreme are the large mainframes that are used, for example, by the banks for performing financial transactions. At the other extreme are the embedded microcomputers which are used to control domestic appliances such as washing machines. Between these extremes are the personal computers and the more powerful engineering workstations. All stages in the production of electronic systems now involve computers, from network design, through preparation of schematics, printed-circuit layout and stock control, to automated manufacturing. Only the first of these applications, computer-aided network analysis and design, will be discussed in this book.

It is often possible to make use of general-purpose network analysis programs, several of which are available commercially. Sometimes, however, special-purpose programs must be written to solve some particular analysis or design problem. In either case it is important to understand the limitations of the underlying analysis method. It is the purpose of this book to describe the most important analysis techniques, and to show how, under some circumstances, they may give misleading or erroneous results.

Electronic network design is a creative process, relying on a mixture of experience and inspiration. Although computers have a perfect memory, and can to some extent be programmed to make use of the experience of a human designer, this is not enough. Engineers face a continual challenge to improve performance in a number of respects, such as power consumption, size, noise and linearity. To meet this challenge requires more than an exhaustive knowledge of previous designs. New network configurations need to be developed, but there is no known systematic way of doing this and progress depends on the inspiration of the designer. The nearest a computer can get to inspiration is to evaluate the performance of randomly chosen network configurations. Whilst this may be practicable for simple networks containing three or four elements, the number of possible configurations for larger networks rapidly exceeds the capabilities of even the most powerful computers.

For this reason computers are normally used indirectly during the design process, to analyze a network at various stages. The design proceeds iteratively. An initial design is produced, often with the aid of simplifying assumptions such as the infinite-gain approximation. A computer is then used to analyze the performance and the results compared with the required specification. In the light of this computer analysis, the design is modified and subjected to further analysis. Eventually, it is to be hoped, a satisfactory design will be obtained.

Traditionally, network designs have been tested by building a prototype. This is an expensive and time-consuming procedure, particularly if the design is to be produced as an integrated circuit. It is of course possible to build a discrete prototype of an integrated circuit, but discrete devices behave differently from integrated devices. A further objection to relying on the performance of a single prototype is that component values vary because of manufacturing tolerances and the effects of ageing and temperature changes. The fact that a prototype built using particular component values meets a specification is no guarantee of an acceptable manufacturing yield.

Nowadays with the availability of powerful computer network analysis and simulation programs it is often possible to bypass the prototype stage altogether. Instead of building a physical prototype the designer determines the performance from a computer model. This is normally quicker and cheaper than the traditional approach. The analysis or simulation can be repeated with different combinations of component values to give the effect of component tolerances.

Most analogue networks are linear, and any non-linear devices such as bipolar or field-effect transistors operate in a quasi-linear mode over a small part of their characteristics. The first stage in analyzing a network is therefore to establish the operating points of any non-linear devices. This involves the solution of simultaneous non-linear equations. Once the operating points are known the non-linear devices can be replaced by their linear small-signal equivalent circuits. Linear analysis techniques can then be used to determine the time-domain or frequency-domain response.

Some networks, such as oscillators, are essentially non-linear and cannot be analyzed using linear methods. To determine the behaviour of such networks it is necessary to perform a non-linear transient analysis. In effect the computer is used to simulate the network. Simulations may be very slow, even on large computers, and several hours may be required to simulate the behaviour of, say, a logic element over a period of 10 ns.

Although computers are normally used to analyze networks, there are some situations where they can be used more directly in the design process. Formal design procedures exist for a limited range of design problems such as filters. If a bandpass filter is required with specified passband and stopband edge frequencies, maximum passband attenuation and minimum stopband attenuation, then a suitable transfer function can be obtained using, say, the elliptic approximation. A passive equally-terminated ladder filter, or its active equivalent, can then be used to implement this filter. There is no reason why the whole of this design process should not be performed by a suitable computer program without human intervention.

Another direct use of computers in design is network optimization. This is a process by which the component values in a network are chosen to give the best performance according to some criterion. In effect the computer minimizes the difference between the specification and the actual performance. Only the component values are determined by the computer, however; the more difficult task of choosing a suitable network configuration is still left to the designer.

1.3 Numerical Precision

Engineering calculations often involve quantities whose values can vary over a wide range. An obvious example is frequency, which, for lumped-element networks, usually lies between 1 Hz and 1 GHz. The range is further increased when such quantities are multiplied or divided by one another. This is a particular problem in frequency response calculations where the frequency must be raised to a power equal to the order of the system. Calculation of the gain of a 12th-order filter from 1 Hz to 10 kHz therefore involves values covering a range of 10^{48}.

Numbers can be stored and processed in a computer using either fixed-point or floating-point representations. Fixed-point numbers use either natural binary or signed binary code together with an implicit scale factor. For example, a 16-bit unsigned fixed-point number with a scale factor of $0 \cdot 01$ can take on values from $0 \cdot 00$ to $655 \cdot 35$. Integer and cardinal numbers are fixed-point representations with unity scale factors. The ratio of the largest to the smallest non-zero number that can be represented in fixed-point form depends only on the number of bits and is independent of the scale factor. For most engineering purposes the range of fixed-point numbers is inadequate. Even a 64-bit fixed-point number only has a range of around 2×10^{19}.

Floating-point numbers consist of two fields, the mantissa M and the exponent E, with the value X being given by

$$X = M \times 2^{E}$$

This is simply the binary equivalent of the familiar decimal scientific notation. Clearly the relative precision of a floating-point number is determined by the number of bits in the mantissa, and its range is determined by the number of bits in the exponent. In the past a variety of floating-point formats have been used by different computer manufacturers. However in 1985 the Institute of Electrical and Electronic Engineers (IEEE) produced a standard for floating-point numbers and this has since been widely adopted.

The IEEE standard defines the format of single (32-bit), double (64-bit) and extended (80-bit) precision floating-point numbers. For the purposes of engineering calculations the double precision format is probably the best choice. It offers a satisfactory precision and range, while making less demands on storage and processing power than the extended format.

Most language compilers for personal computers use the IEEE double precision format for real numbers. Figure 1.1 shows the structure of IEEE double precision numbers. The mantissa is a 53-bit unsigned binary number, consisting of a most significant hidden bit, whose value is always 1, together with the 52 bits of the IEEE format. Immediately to the right of the hidden bit is the binary point. The mantissa can therefore vary from

$1 \cdot 000 \ldots 000 = 1$ (decimal)

Figure 1.1
The IEEE
double
precision
format

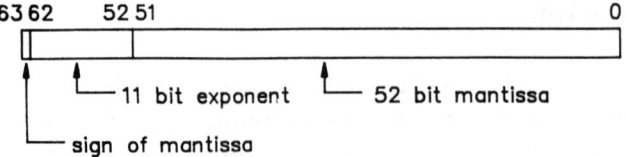

to

$$1 \cdot 111 \ldots 111 \approx 2 \quad \text{(decimal)}$$

A change in the least significant bit is equal to $2^{-52} = 2 \cdot 2 \times 10^{-16}$; since the smallest value of the mantissa is unity then $2 \cdot 2 \times 10^{-16}$ is also the relative precision. In other words an IEEE double precision number has between 15 and 16 significant decimal digits.

The exponent is an 11-bit excess 1023 binary number, except that the bit patterns

00000000000

and

11111111111

are reserved for special cases. Exponent values therefore range from

$$00000000001 = -1022 \text{ (decimal)}$$

to

$$11111111110 = 1023 \text{ (decimal)}$$

Combining exponent and mantissa gives an overall range from

$$1 \times 2^{-1022} = 2 \cdot 226 \times 10^{-308}$$

to

$$2 \times 2^{1023} = 1 \cdot 797 \times 10^{308}$$

Normally the results of engineering calculations are required to an accuracy of at most six decimal digits. It might therefore be supposed that a real number precision of 15 digits would be more than adequate, but this is not necessarily so. There are several ways in which a loss of precision can occur, some of which may render the result of a calculation quite meaningless.

When a small number is added to a larger number, a rounding error related to the precision of the larger number occurs. If this procedure is repeated a large number of times then the effects of the rounding errors can become quite significant. For example when a value of 10^{-15} is added to $1 \cdot 0$ a total of one million times the following result is obtained:

$$1 \cdot 00000000111022$$

In spite of the fact that the operations are performed using 15-digit precision, the result is accurate only to 9 digits.

A more serious loss of precision occurs when quantities of similar magnitude are subtracted. The absolute precision of the result is equal to the absolute precision of the subtracted quantities. If the result is smaller in magnitude than the subtracted quantities then it will have a reduced relative precision. For example, when $1000000 \cdot 0$ is subtracted from $1000000 \cdot 1$ using IEEE double precision real numbers the result is

$$0 \cdot 0999999999767169$$

and six digits of accuracy have been lost.

Consider evaluation of the simple formula

$$X = (A + B) - B \tag{1.3.1}$$

Obviously the answer should be A, but if B is larger in magnitude, and of the same sign as A, then A will suffer a loss in relative precision. In an extreme case all information about A may be lost. For example the result of evaluating the formula with $A = 0 \cdot 1$ and $B = 10^{15}$ using IEEE double precision real numbers is

$$0.125000000000000$$

Operations, similar to that expressed in the formula, are implicit in many engineering calculations. One example is in the solution of simultaneous linear equations. If the equations are poorly conditioned (that is the matrix of coefficients is nearly singular) then subtraction of nearly-equal quantities may seriously degrade the solution. Another example concerns the use of exponential and logarithm functions. If A is small then

$$X = \exp(A) \approx 1 \cdot 0 + A \tag{1.3.2}$$

Clearly this operation reduces the information about A in the result X. Taking the logarithm of X returns the original value of A, but at reduced precision.

These examples should serve as a warning that care is needed when interpreting the results of engineering calculations. In some circumstances the accuracy may fall well below that of the real number representation.

1.4 The Language Modula-2

In 1980 Niklaus Wirth published a description of a new computer language called Modula-2. Since then Modula-2 has gained widespread acceptance as a simple but powerful general-purpose computer language. Modula-2 has many features in common with the language Pascal which was also designed by Wirth. In particular Modula-2 has strong typing, a rich variety of data types and powerful control statements.

The origins of Modula-2 can be traced back to the early programming language Algol-60. Designed by an international committee and published in 1960, Algol-60 was very different from existing high-level languages such as Fortran. Algol-60 was the first block-structured computer language, the first language to have a formal syntactic definition, and the first to distinguish between local and global variables. Although it became popular in Europe, Algol-60 never challenged the supremacy of Fortran in the United States. Nevertheless, Algol-60 has had a profound influence on the subsequent development of computer languages.

In the years following the introduction of Algol-60 a number of attempts were made to create improved versions of the language. These culminated in 1968 with the definition of a new language, Algol-68. Unfortunately, although it is both elegant and powerful, Algol-68 is difficult to implement and is not particularly suitable as an initial teaching language. Wirth believed that there was a requirement for a simpler language that would support structured programming and that could be efficiently implemented. In 1971 he introduced the language Pascal.

Pascal rapidly gained widespread acceptance and its use spread from the universities into industry. One of the reasons for its success is that the one-pass compiler is itself written in Pascal, and can readily be transferred to new computer hardware. As a result, in spite of receiving little support from computer manufacturers, Pascal quickly became available on most computers.

Wirth had never intended Pascal to be used for large-scale software development projects involving many programmers. He recognized that software complexity increases rapidly with size and that a single large program containing, say, 50 000 lines of code is much easier to develop and maintain if it can be split into 25 modules, each of 2000 lines. In response to this need Wirth designed a new language, Modula-2, and published a description of this language in 1980.

Modula-2 is clearly based on Pascal, but has been extended in a number of ways while retaining the simplicity, clarity and compactness of the older language. The primary new feature of Modula-2 is its support for separate compilation. A Modula-2 program consists of a program module together with a number of library modules. Each library module has two parts, a definition part and an implementation part, and this allows rigorous type checking between modules during compilation. Library modules usually consist of a number of related entities (constants, types, variables and procedures), but not all of these are visible outside the modules. Only entities that have been specifically exported from a module are available for use by other modules.

Modula-2 supports a form of concurrent programming using coroutines, and it also provides low-level machine access. These features make it a suitable language for programming embedded computer systems.

Of course, Modula-2 is not the only language to support structured programming. At about the same time that Wirth was developing Modula-2, an international design team was working on a new software implementation language for the American Department of Defence. Ada, as this language became known, was finally released in 1983. Ada is an extremely powerful language and in some respects, such as generics, exception handling and concurrency, is superior to Modula-2. It is also, however, a large and complex language. As a result Ada compilers tend to be rather slow and to make considerable demands on memory capacity both during compilation and at run-time.

At the present time (1990) Ada is not so widely available as Modula-2, and existing Ada compilers tend to be more expensive and less convenient to use than Modula-2 compilers. Because of its complexity Ada is not particularly suitable for use as an initial teaching language. For these reasons Modula-2 was selected to illustrate the various algorithms discussed in this book.

There are several excellent Modula-2 compilers available for IBM personal computers. All of the examples given in this book were tested using the Logitek Modula-2/86 system. Whilst it does not compile programs as rapidly as, for example, JPI Topspeed Modula-2, Logitek Modula-2/86 has superior debugging facilities. These include symbolic post-mortem and run-time debuggers. In practice the time saved during program development by the use of these Logitek debuggers more than compensates for the lower compilation speed.

2 Frequency-domain Response of Linear Networks

2.1 Introduction

The response of a linear electrical network in the time domain is characterised by its impulse response $h(t)$ (see figure 2.1). $h(t)$ is defined to

$$\underset{\text{Input}}{\overset{x(t)}{\longrightarrow}} \boxed{h(t)} \underset{\text{Output}}{\overset{y(t)}{\longrightarrow}}$$

be the output of the network that would result from a unit impulse $\delta(t)$ applied to the input where

$$\delta(t) = 0 \quad \text{for} \quad t \neq 0$$
$$\int_{-\infty}^{\infty} \delta(t)\ \mathrm{d}t = 1 \tag{2.1.1}$$

Such an input is not, of course, physically realizable, and to measure the impulse response of a network it is necessary to approximate $\delta(t)$ by a pulse of finite amplitude and finite duration. For example a rectangular pulse might be used:

$$\begin{aligned} x(t) &= x_0 \quad \text{for} \quad 0 \leqslant t < \Delta\tau \\ &= 0 \quad \text{for} \quad t < 0 \quad \text{and} \quad t \geqslant \Delta\tau \end{aligned} \tag{2.1.2}$$

Provided that the amplitude x_0 is insufficient to drive the network into non-linearity, and provided that the duration $\Delta\tau$ is much shorter than any time constants of the network, then the output $y(t)$ provides a good approximation to the impulse response:

$$h(t) = \frac{1}{x_0\ \Delta\tau}\ y(t) \tag{2.1.3}$$

The response of the network to a general input $x(t)$ can be derived by

Figure 2.2
A general input
waveform

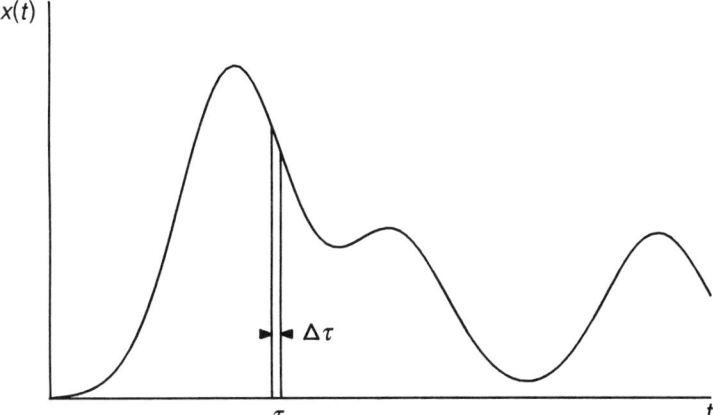

considering the input waveform to be composed of a large number of rectangular pulses. A pulse centred on a time τ and of amplitude $x(\tau)$ (figure 2.2) will give rise to an output

$$y(t) = x(\tau) \, \Delta\tau \, h(t - \tau) \tag{2.1.4}$$

Linearity of the network implies that the output resulting from all of the pulses is equal to the sum of the responses to the individual pulses:

$$y(t) = \sum_{i=-\infty}^{\infty} x(\tau_i) \, \Delta\tau \, h(t - \tau_i) \tag{2.1.5}$$

where $\tau_i = i \, \Delta\tau$.

The accuracy of this equation improves as the pulse width $\Delta\tau$ is reduced and in the limit as $\Delta\tau \to 0$ the response becomes

$$y(t) = \int_{-\infty}^{\infty} x(\tau)h(t - \tau) \, d\tau \tag{2.1.6}$$

This equation shows that the output of a linear network is the convolution of its impulse response with the input. However, it is not in practice a particularly convenient way of expressing the response of the network. Not only must a differential equation be solved to obtain $h(t)$, but for each value of t at which the output is to be determined, an integral must be evaluated.

The Laplace transform method is commonly used in order to avoid these difficulties. A function of time, $f(t)$, has a Laplace transform $F(s)$, where

$$F(s) = \mathscr{L}f(t) = \int_{0-}^{\infty} f(t)e^{-st} \, dt \tag{2.1.7}$$

and $s = \sigma + j\omega$ is complex

$F(s)$ does not depend on t because the integral has fixed limits. The lower integration limit is specified as $0-$, rather than 0 or $0+$, in order to include cases where the function $f(t)$ is discontinuous in value at $t = 0$. If $X(s)$ and

11

$Y(s)$ are the Laplace transforms of the input and output of the network:

$$X(s) = \mathcal{L}x(t)$$
$$Y(s) = \mathcal{L}y(t)$$
(2.1.8)

then the response of the network can be expressed in the new form:

$$Y(s) = \mathcal{L} \int_{-\infty}^{\infty} x(\tau)h(t - \tau)\, d\tau$$
(2.1.9)

Provided that $h(t) = 0$ for all $t < 0$, that is to say that the system is causal, and that the input $x(t) = 0$ for all $t < 0$, equation (2.1.9) becomes

$$Y(s) = \int_{0-}^{\infty} h(t)e^{-st}\, dt \int_{0-}^{\infty} x(t)e^{-st}\, dt$$

$$= H(s)X(s)$$
(2.1.10)

where

$$H(s) = \mathcal{L}h(t)$$
(2.1.11)

$H(s)$, the Laplace transform of the impulse response, is known as the transfer function (figure 2.3). In most cases $H(s)$ will be a voltage transfer function, although in principle it can relate the voltage or current response

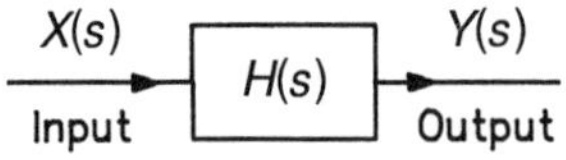

Figure 2.3
The transfer
function

at any point in the network to a voltage or current applied at any point in the network.

A network containing no distributed elements (such as delay lines) will have a transfer function which is the ratio of two polynomials in s:

$$H(s) = \frac{N(s)}{D(s)}$$

$$= \frac{a_0 + a_1 s + a_2 s^2 + \cdots + a_n s^n}{b_0 + b_1 s + b_2 s^2 + \cdots + b_n s^n}$$
(2.1.12)

where the coefficients $a_0 \ldots a_n$, $b_0 \ldots b_n$ are real. It follows that $H(s)$ has conjugate symmetry, that is

$$H(s\star) = H\star(s)$$
(2.1.13)

The advantages of using the Laplace transform representation are obvious. Simple multiplication of the input by the transfer function is all that is required to obtain the output. Moreover, the transfer function can easily be determined by using the fact that

$$\mathcal{L}\frac{df}{dt} = s\mathcal{L}f(t) - f(0-)$$
(2.1.14)

12

Applying this identity to the defining equations for capacitors and inductors:

$$v(t) = L \frac{\mathrm{d}i(t)}{\mathrm{d}t}$$

$$i(t) = C \frac{\mathrm{d}v(t)}{\mathrm{d}t}$$

(2.1.15)

gives

$$V(s) = sLI(s) - Li(0-)$$
$$I(s) = sCV(s) - Cv(0-)$$

(2.1.16)

Assuming that the currents in the inductors and the voltages across the capacitors are zero at $t = 0$, allows these equations to be simplified, giving the Laplacian impedances Z_L and Z_C:

$$Z_L(s) = \frac{V(s)}{I(s)} = sL$$

$$Z_C(s) = \frac{V(s)}{I(s)} = \frac{1}{sC}$$

(2.1.17)

Analysis of the network using the Laplacian impedances leads directly to the transfer function $H(s)$.

If the time-domain response to a particular input is required, then it can in principle be derived by performing an inverse Laplace transform on $Y(s)$:

$$y(t) = \mathscr{L}^{-1} Y(s)$$

$$= \frac{1}{2\pi \mathrm{j}} \int_{\sigma - \mathrm{j}\infty}^{\sigma + \mathrm{j}\infty} Y(s) \mathrm{e}^{st} \, \mathrm{d}s$$

(2.1.18)

Methods for calculating the time-domain behaviour of linear networks are discussed in detail in chapter 6.

2.2 The Frequency Response Function

An input function which is of particular importance in the context of linear networks is the complex exponential:

$$x(t) = x_0 \mathrm{e}^{st}$$

(2.2.1)

Substituting this input function into equation (2.1.6) gives the response of the network:

$$y(t) = \int_{-\infty}^{\infty} x_0 \mathrm{e}^{s\tau} h(t - \tau) \, \mathrm{d}\tau$$

(2.2.2)

$$= H(s) x_0 \mathrm{e}^{st}$$

(2.2.3)

The response to a complex exponential input is a complex exponential output. In other words, the complex exponential is an eigenfunction of the linear operation performed by the network; the transfer function $H(s)$ is the corresponding eigenvalue.

Writing the complex quantity s as the sum of its real and imaginary parts $s = \sigma + j\omega$ shows that the complex exponential input function in fact represents an exponentially increasing, or decreasing, sinusoid:

$$x(t) = x_0 e^{st}$$

$$= x_0 e^{\sigma t} e^{j\omega t}$$

$$= x_0 e^{\sigma t} (\cos \omega t + j \sin \omega t) \tag{2.2.4}$$

A steady-state input can be obtained by setting σ, the real part of s, to zero so that

$$s = j\omega$$

$$x(t) = x_0 e^{j\omega t} \tag{2.2.5}$$

and the response of the network to this input will be given by

$$y(t) = H(j\omega) x_0 e^{j\omega t} \tag{2.2.6}$$

$H(j\omega)$ is known as the frequency response function and is obtained from the transfer function simply by substituting $j\omega$ in place of s.

The input function given in equation (2.2.5) is complex and is not, of course, physically realizable. However, the closely related cosine input is physically realizable, and can be considered to be the sum of two complex exponential inputs:

$$x(t) = x_0 \cos \omega t \tag{2.2.7}$$

$$= \frac{x_0}{2} (e^{j\omega t} + e^{-j\omega t}) \tag{2.2.8}$$

Linearity of the network implies that the response is equal to the sum of the responses to the individual complex exponentials:

$$y(t) = \frac{x_0}{2} H(j\omega) e^{j\omega t} + \frac{x_0}{2} H(-j\omega) e^{-j\omega t} \tag{2.2.9}$$

Now $H(s)$ possesses conjugate symmetry (see equation 2.1.13) so that

$$H(-j\omega) = H^\star(j\omega) \tag{2.2.10}$$

and the expression for the output simplifies to

$$y(t) = \frac{x_0}{2} \{H(j\omega) e^{j\omega t} + H^\star(j\omega) e^{-j\omega t}\}$$

$$= x_0 \, \text{Re}\{H(j\omega) e^{j\omega t}\} \tag{2.2.11}$$

If $H(j\omega)$ is written in polar form:

$$H(j\omega) = A(\omega) e^{j\theta(\omega)} \tag{2.2.12}$$

then the response to a cosine input is finally obtained:

$$y(t) = x_0 A(\omega)\cos\{\omega t + \theta(\omega)\} \qquad (2.2.13)$$

From this expression it is seen that a linear network converts a sinusoidal input into a sinusoidal output of the same frequency, but of different amplitude and phase. Furthermore, the gain A is the magnitude of $H(j\omega)$, and the phase shift θ is the angle of $H(j\omega)$. Both A and θ are, in general, dependent on the frequency ω of the sinusoid.

2.3 Frequency Response Evaluation

The network gain and phase can both be obtained from the frequency response function $H(j\omega)$:

$$H(j\omega) = \frac{N(j\omega)}{D(j\omega)}$$

$$= \frac{a_0 + a_1 j\omega + a_2(j\omega)^2 + \cdots + a_n(j\omega)^n}{b_0 + b_1 j\omega + b_2(j\omega)^2 + \cdots + b_n(j\omega)^n} \qquad (2.3.1)$$

In this expression the polynomial coefficients $a_0 \ldots a_n$, $b_0 \ldots b_n$ are real and depend on the component values of the network. Although the numerator and denominator order are equal this does not limit the generality of the expression; if the numerator order is in fact less than the denominator order, then some of the coefficients a_n, $a_{n-1}, \ldots$ can be set to zero.

The first stage in determining $H(j\omega)$ is the evaluation of the numerator and denominator polynomials. Complex division of $N(j\omega)$ by $D(j\omega)$ then gives $H(j\omega)$. Consider then the problem of evaluating a polynomial in $j\omega$:

$$N(j\omega) = a_0 + a_1 j\omega + a_2(j\omega)^2 + \cdots + a_n(j\omega)^n \qquad (2.3.2)$$

At first sight it might appear that a term $a_r(j\omega)^r$ in the polynomial requires r multiplications for its evaluation, so that calculation of the value of the complete polynomial involves a total of $n(n+1)/2$ multiplications. However, $(j\omega)^r$ can be obtained from $(j\omega)^{r-1}$, which is used in the previous term, by a single multiplication. Consequently two multiplications are necessary for each term, and $2n$ multiplications are required to evaluate the complete polynomial.

A more efficient method of polynomial evaluation becomes apparent if the polynomial is rearranged in nested form:

$$N(j\omega) = a_0 + j\omega(a_1 + j\omega(a_2 + \cdots + j\omega a_n))\ldots)) \qquad (2.3.3)$$

Only one multiplication is required for each power of $j\omega$, and n multiplications are necessary to evaluate the complete polynomial. This procedure, which is known as Horner's rule, can be summarized by the recursion formula:

$$N(j\omega) = p_0$$
$$p_n = a_n \qquad\qquad\qquad\qquad (2.3.4)$$
$$p_{n-k} = a_{n-k} + j\omega p_{n-k+1} \qquad k = 1, 2, \ldots, n$$

Unfortunately, although this procedure is quite satisfactory for values of ω close to and less than unity, it can result in numeric overflow when a computer is used to evaluate a high-order frequency response function at large values of ω. To take an extreme case, consider a 20th-order high-pass filter. The coefficients a_{20} and b_{20} will be of similar size, and the response will tend to a value a_{20}/b_{20} at high frequencies. However, the magnitudes of the numerator and denominator polynomials will tend to $a_{20}\omega^{20}$ and $b_{20}\omega^{20}$ respectively, and these values may exceed the floating-point range of the computer.

A simple way of avoiding this difficulty is to divide both numerator and denominator polynomials by $(j\omega)^n$, thus leaving the value of $H(j\omega)$ unchanged:

$$H(j\omega) = \frac{N'(j\omega)}{D'(j\omega)}$$

$$= \frac{a_0/(j\omega)^n + a_1/(j\omega)^{n-1} + \cdots + a_n}{b_0/(j\omega)^n + b_1/(j\omega)^{n-1} + \cdots + b_n} \qquad (2.3.5)$$

The polynomials can be arranged in nested form, for example:

$$N'(j\omega) = a_n + \frac{1}{j\omega}\left(a_{n-1} + \frac{1}{j\omega}\left(a_{n-2} + \cdots + \frac{a_0}{j\omega}\right)\ldots\right) \qquad (2.3.6)$$

$$= a_n - \frac{j}{\omega}\left(a_{n-1} - \frac{j}{\omega}\left(a_{n-2} - \cdots - j\,\frac{a_0}{\omega}\right)\ldots\right) \qquad (2.3.7)$$

Horner's rule can now be used to evaluate the modified polynomials:

$$N'(j\omega) = p_n$$
$$p_0 = a_0 \qquad\qquad\qquad\qquad (2.3.8)$$
$$p_k = a_k - j p_{k-1}/\omega \qquad k = 1, 2, \ldots n$$

At high frequencies the numerator polynomial tends to a value of a_n and the denominator polynomial tends to a value of b_n. There is no danger of numeric overflow. However, this method will fail at very low frequencies. In particular, a value of $\omega = 0$, where $H(j\omega) = a_0/b_0$, will give divide-by-zero errors during polynomial evaluation.

Since neither of the two methods described will work correctly for all frequencies, a decision must be made, each time $H(j\omega)$ is to be determined, on which procedure to adopt. Both methods are satisfactory for a considerable range of frequencies around $\omega = 1$ and difficulties are only encountered in the frequency extremes. A suitable criterion might be whether ω is greater or less than unity. For $\omega \leqslant 1.0$ the formula (2.3.4) should be used; for $\omega > 1.0$ the formula (2.3.8) should be used.

Modula-2 is a general-purpose computer language and does not include complex numbers amongst its simple data types. However, a new type *complex* can be declared as a record consisting of two real variables:

```
TYPE
    complex = RECORD re, im: REAL END;
```

It would, in principle, be possible to create a module containing a package of procedures to manipulate complex variables. Such a package might include procedures to add, subtract, multiply and divide complex quantities. Unfortunately, Modula-2 does not permit definition or overloading of operators; nor does it permit functions to return record types. The only mechanism for returning a record provided by Modula-2 is the *VAR* procedure parameter. That being the case it is probably less cumbersome, and certainly more efficient, to write out explicitly the operations on the real and imaginary parts of the complex quantities, rather than to use procedures.

The frequency response function can be represented by a record consisting of two polynomials and a cardinal. Since the polynomial coefficients are real, the polynomials can themselves be represented by real arrays.

```
TYPE
    index = [0..maxsize];
    poly = ARRAY index OF REAL;
    rational = RECORD a, b: poly;
                       order: index
                 END;
```

A procedure *eval* which calculates the complex value q of the frequency response function H at a frequency *omega* is shown below:

```
PROCEDURE eval(H: rational; omega: REAL; VAR q: complex);
VAR
    j: CARDINAL;
    dr, nr, di, ni, dt, nt: REAL;
BEGIN
    WITH H DO
        IF omega > 1.0 THEN
            nr := a[0]; ni := 0.0;
            dr := b[0]; di := 0.0;
            FOR j := 1 TO order DO
                nt := a[j]+ni/omega; ni := -nr/omega; nr := nt;
                dt := b[j]+di/omega; di := -dr/omega; dr := dt
            END
        ELSE
            nr := a[order]; ni := 0.0;
            dr := b[order]; di := 0.0;
            FOR j := order-1 TO 0 BY -1 DO
                nt := a[j]-ni*omega; ni := nr*omega; nr := nt;
                dt := b[j]-di*omega; di := dr*omega; dr := dt
            END
        END
    END;
    dt := dr*dr+di*di;
    q.re := (nr*dr+ni*di)/dt;
    q.im := (ni*dr-nr*di)/dt
END eval;
```

2.4 The Network Gain

The gain A of a network is given by the modulus of its frequency response function:

$$A = |H(j\omega)| = \sqrt{(x^2 + y^2)} \tag{2.4.1}$$

where $H(j\omega) = x + jy$.

At first sight this appears to be a straightforward operation which could be performed by the procedure shown below:

```
PROCEDURE gain(H: rational; omega: REAL): REAL;
VAR
    q: complex;
BEGIN
    eval(H, omega, q);
    RETURN sqrt(q.re*q.re+q.im*q.im)
END gain;
```

However, since the result is derived directly from a square root, the range of possible values for the gain is limited to half the floating-point range. A more satisfactory procedure is shown below:

```
PROCEDURE gain(H: rational; omega: REAL): REAL;
VAR
    q: complex;
    gr, gi: REAL;
BEGIN
    eval(H, omega, q);
    gr := ABS(q.re); gi := ABS(q.im);
    IF gr = 0.0 THEN
        RETURN gi
    ELSIF gr > gi THEN
        RETURN gr*sqrt(1.0+(gi/gr)*(gi/gr))
    ELSE
        RETURN gi*sqrt(1.0+(gr/gi)*(gr/gi))
    END
END gain;
```

Conversion of the gain into decibels (dB) is performed using the formula:

$$\text{Gain(dB)} = 20 \log_{10} A$$

$$= 20 \frac{\ln A}{\ln 10} \qquad (2.4.2)$$

Before this formula can be applied the gain A must be checked in case it happens to be zero. If it is zero then a result of *minus infinity* should be returned; if not then the logarithm can be evaluated safely. (*infinity* is the largest number that can be represented in floating-point form, and will depend on the particular compiler being used.)

2.5 The Network Phase

The phase shift θ of a network is given by the angle of $H(j\omega)$:

$$\tan \theta = y/x \qquad\qquad (2.5.1)$$

where $H(j\omega) = x + jy$.

Although Modula-2 does not provide an inverse tangent function as part of the language, this function will usually be available from a library module (such as, for example, *MathLib0*). The angle returned by the inverse tangent function will lie within a range of π radians, usually $-\pi/2$ to $+\pi/2$. Of course x and y together uniquely define an angle in a 2π range, but their individual signs are lost when y is divided by x. In order to determine θ within the full 2π range, a value is first obtained within the range $-\pi/2$ to $+\pi/2$ from the inverse tangent function:

$$\theta = \arctan y/x \qquad\qquad (2.5.2)$$

This value is then shifted by π if the real part x of $H(j\omega)$ is negative. Care must be taken when evaluating expression (2.5.2) to prevent numeric overflow if the value of x is very small.

A procedure for calculating the network phase is shown below:

```
PROCEDURE phase(H: rational; omega: REAL): REAL;
VAR
    q: complex;
    t: REAL;
BEGIN
    eval(H, omega, q);
    WITH q DO
        IF ABS(im) = ABS(re)+ABS(im) THEN
            IF im < 0.0 THEN RETURN -90.0 ELSE RETURN -270.0 END
        END;
        t := arctan(im/re)*180.0/pi;
        IF re < 0.0 THEN t := t+180.0 END;
        IF t > 0.0 THEN RETURN t-360.0 ELSE RETURN t END
    END
END phase;
```

2.6 Display of the Frequency Response

An important part of frequency response calculations is the presentation of the results. Consider the filter network shown in figure 2.4. Analysis of this

Figure 2.4
A 5th-order
elliptic filter

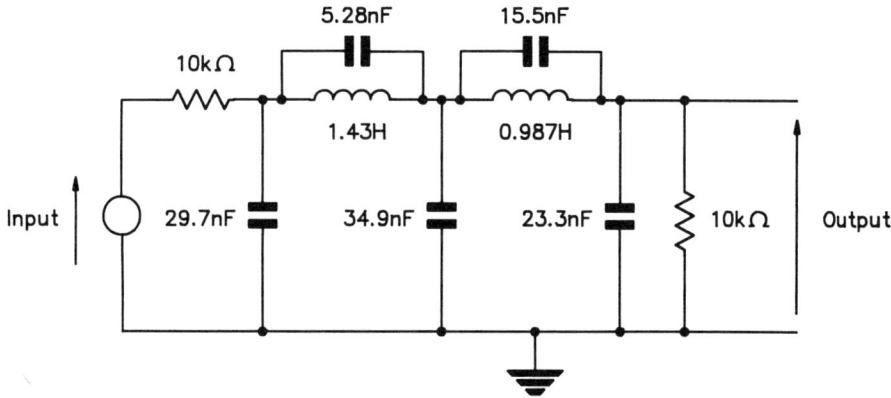

network shows that the frequency response function is of 5th order with coefficients:

$$a_0 = 0\cdot5 \qquad b_0 = 1\cdot0$$
$$a_1 = 0\cdot0 \qquad b_1 = 5\cdot61 \times 10^{-4}$$
$$a_2 = 1\cdot14 \times 10^{-8} \qquad b_2 = 1\cdot29 \times 10^{-7}$$
$$a_3 = 0\cdot0 \qquad b_3 = 3\cdot38 \times 10^{-11}$$
$$a_4 = 5\cdot76 \times 10^{-17} \qquad b_4 = 2\cdot71 \times 10^{-15}$$
$$a_5 = 0\cdot0 \qquad b_5 = 4\cdot66 \times 10^{-19}$$

The gain and phase of this transfer function can now be evaluated at a sequence of frequencies spanning the range of interest and the results presented in numerical form:

Frequency	Gain	Phase
$1\cdot000000E+00$	$-6\cdot02$	$359\cdot97$
$1\cdot000000E+01$	$-6\cdot02$	$359\cdot68$
$1\cdot000000E+02$	$-6\cdot03$	$356\cdot79$
$1\cdot000000E+03$	$-6\cdot40$	$328\cdot89$
$1\cdot000000E+04$	$-51\cdot37$	$129\cdot39$
$1\cdot000000E+05$	$-58\cdot29$	$273\cdot33$
$1\cdot000000E+06$	$-78\cdot17$	$270\cdot33$

Unfortunately, if the frequencies are chosen to be sufficiently close to show up details in the response, then the amount of data will be large and difficult to absorb in this form.

Whenever possible, frequency responses should be presented in graphical form, and the most convenient way of doing this is as a Bode plot. A Bode plot consists of the gain (expressed in dB) and the phase plotted on a logarithmic frequency scale. They may be displayed separately, as shown in figure 2.5, or combined into a single graph.

An alternative way of presenting the frequency response is on a polar plot where the value of $H(j\omega)$ is plotted on the complex plane as the frequency

Figure 2.5
A Bode plot of
a 5th-order
elliptic
response

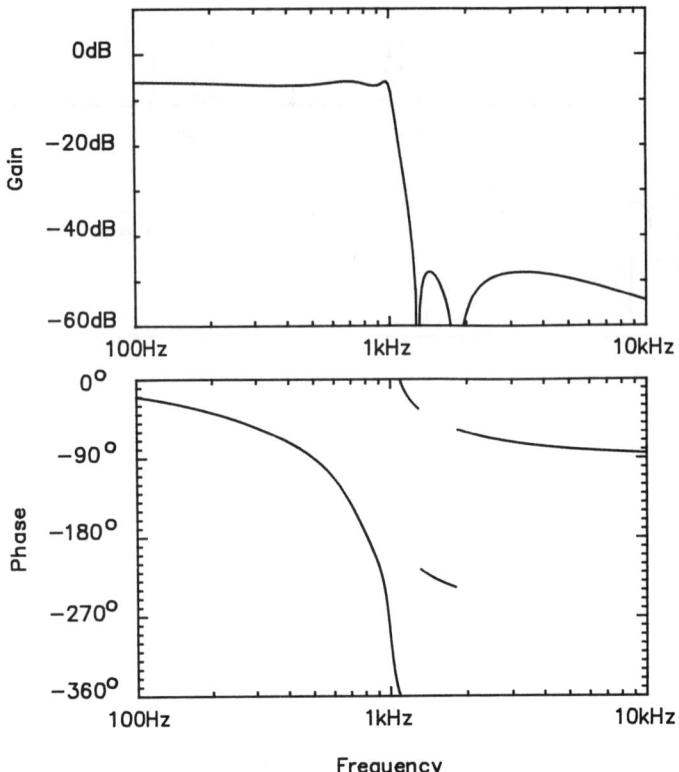

varies from zero to infinity. Polar plots are of particular importance in feedback control system design where a polar plot of the loop gain is known as a Nyquist diagram. A typical polar plot is shown in figure 2.6.

Another useful way of displaying a frequency response is on a Nichol plot. Here the phase is plotted against the gain (expressed in dB) as the frequency varies from zero to infinity. Figure 2.7 shows a typical Nichol plot.

Figure 2.6
A polar plot

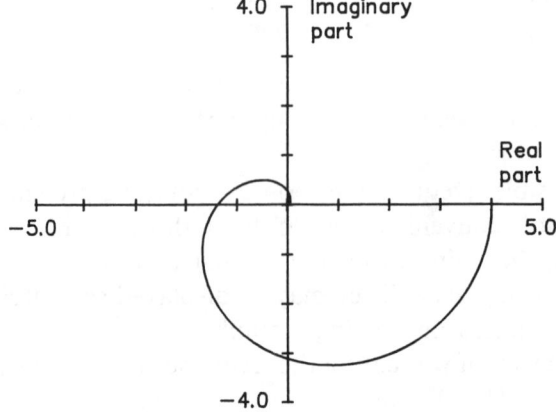

Figure 2.7
A Nichol plot

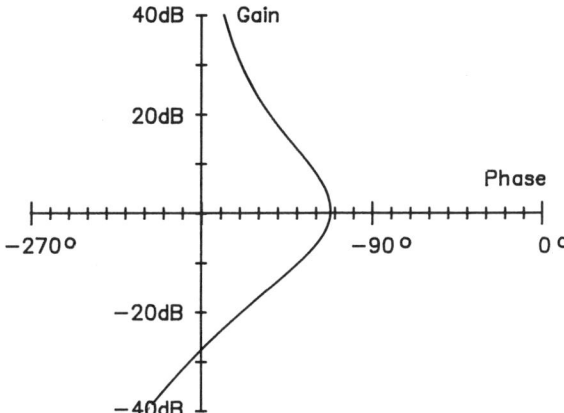

A particular feature of Nichol plots is the ease with which the gain and phase margins can be estimated. (The gain margin is the gain when the phase is $-180°$; the phase margin is the difference between the phase at 0 dB gain, and $-180°$.) Gain and phase margins are important properties of feedback control systems.

2.7 Summary

Linear systems are characterized in the time domain by their impulse response $h(t)$, and the output of a linear system is the convolution of its input with $h(t)$. Laplace transforms are used to avoid the need for convolution. The Laplace transform of $h(t)$ is the transfer function $H(s)$, and the output transform is equal to the product of $H(s)$ and the input transform. Networks with no distributed elements (such as delay lines) have rational transfer functions with real numerator and denominator polynomial coefficients.

Complex exponential signals of the form $e^{j\omega t}$ are of particular importance: the response of a linear system to a complex exponential input is a complex exponential output. Such signals are not, of course, physically realizable, but a real sinusoid can be regarded as the sum of two complex exponentials. The response of a linear system to a real sinusoidal input is a real sinusoidal output of different amplitude and phase.

The gain of a linear system at a frequency ω is given by the modulus of the frequency response function $H(j\omega)$, and the phase shift is given by the angle of $H(j\omega)$. A Horner evaluation method can be used to calculate the complex values of the numerator and denominator polynomials $N(j\omega)$ and $D(j\omega)$. Dividing $N(j\omega)$ by $D(j\omega)$ then gives $H(j\omega)$. Unfortunately, to prevent overflow during polynomial evaluation it is necessary to use different methods for $\omega < 1$ and $\omega \geqslant 1$.

Frequency responses should normally be displayed in graphical rather than numerical form. In the case of filters or similar signal processing networks the appropriate form of display is a Bode diagram which shows the variation of gain and phase with frequency. Nyquist and Nichol diagrams are used in control system design.

2.8 Problems

1 Notch filters have frequency response functions of the general form:

$$H(j\omega) = \frac{1 + (j\omega)^2/\omega_0^2}{1 + j\omega/(\omega_0 Q) + (j\omega)^2/\omega_0^2}$$

where ω_0 is the notch frequency and Q controls the sharpness of the notch. Obtain Bode diagrams showing the gain in dB over the frequency range 100 Hz to 10 kHz of notch filters with $\omega_0 = 2000\pi$ and $Q = 0\cdot5$, 1, 2, 5.

2 The frequency response functions of 5th-order Butterworth, Chebychev and elliptic low-pass filters are given below:

$$H_B(j\omega) = \frac{9180}{9180 + 5050j\omega + 1300(j\omega)^2 + 207(j\omega)^3 + 20\cdot3(j\omega)^4 + (j\omega)^5}$$

$$H_C(j\omega) = \frac{613}{613 + 636j\omega + 136(j\omega)^2 + 55\cdot9(j\omega)^3 + 3\cdot61(j\omega)^4 + (j\omega)^5}$$

$$H_e(j\omega) = \frac{1060 + 23\cdot1(j\omega)^2 + 0\cdot111(j\omega)^4}{1060 + 856j\omega + 157(j\omega)^2 + 61\cdot6(j\omega)^3 + 3\cdot58(j\omega)^4 + (j\omega)^5}$$

Obtain Bode diagrams for these filters showing the gain in dB and the phase over the frequency range $0\cdot1$ Hz to 10 Hz.

3 The frequency response function of a 5th-order Bessel filter is given by

$$H(j\omega) = \frac{945}{945 + 945j\omega + 420(j\omega)^2 + 105(j\omega)^3 + 15(j\omega)^4 + (j\omega)^5}$$

Obtain a Bode diagram for this filter showing the gain in dB, the phase ϕ, and the group delay $(d\phi/d\omega)$ over the frequency range $0\cdot1$ Hz to 10 Hz.

4 Two all-pass filters are used as part of a single-sideband modulator to generate a differential phase shift of $90°$ over the frequency range 50 Hz to 5 kHz. The frequency response functions $H_1(j\omega)$, $H_2(j\omega)$ of the filters are given by

$$H_1(j\omega) = \frac{(-2\cdot10 \times 10^{11}) + (3\cdot57 \times 10^8 j\omega) - (6\cdot62 \times 10^4 (j\omega)^2) + (j\omega)^3}{(2\cdot10 \times 10^{11}) + (3\cdot57 \times 10^8 j\omega) + (6\cdot62 \times 10^4 (j\omega)^2) + (j\omega)^3}$$

$$H_2(j\omega) = \frac{(-4\cdot58 \times 10^9) + (3\cdot07 \times 10^7 j\omega) - (1.68 \times 10^4 (j\omega)^2) + (j\omega)^3}{(4\cdot58 \times 10^9) + (3\cdot07 \times 10^7 j\omega) + (1\cdot68 \times 10^4 (j\omega)^2) + (j\omega)^3}$$

Evaluate the difference in phase generated by these filters over the frequency range 10 Hz to 25 kHz. The phase error is the amount by which the differential phase shift departs from 90°. Determine the maximum phase error over the frequency range 50 Hz to 5 kHz.

5   A control system has a loop gain given by

$$H(j\omega) = \frac{1 + (2 \times 10^{-2}j\omega)}{(2 \times 10^{-4}(j\omega)^2) + (1 \cdot 25 \times 10^{-6}(j\omega)^3) + (1 \cdot 25 \times 10^{-9}(j\omega)^4)}$$

Obtain Nyquist and Nichol plots for this control system over the frequency range 1 Hz to 10 kHz. Use the Nichol plot to estimate the gain and phase margins of the control system.

3 Poles and Zeros of the Transfer Function

3.1 Introduction

Linear networks containing no distributed elements have transfer functions which are the ratio of two polynomials in s. The numerator and denominator polynomials can be factorized:

$$H(s) = \frac{N(s)}{D(s)}$$

$$= G \frac{(s - z_1)(s - z_2)...(s - z_n)}{(s - p_1)(s - p_2)...(s - p_n)}$$

(3.1.1)

Zero values of the numerator polynomial, which occur at $s = z_1$, $s = z_2, ... s = z_n$, lead to $H(s)$ being zero, and $z_1, z_2, ... z_n$ are known as the *zeros* of the transfer function. Zero values of the denominator polynomial, which occur at $s = p_1, s = p_2, ... s = p_n$, make $H(s)$ infinite, and $p_1, p_2, ... p_n$ are known as the *poles* of the transfer function. Both poles and zeros are, in general, complex. Since the coefficients of the numerator and denominator polynomials are real, the poles and zeros, if not themselves real, must occur in complex conjugate pairs.

The poles, and to a lesser extent the zeros, of a transfer function determine its transient response. $H(s)$ is the Laplace transform of the impulse response $h(t)$; the impulse response can therefore be found by applying an inverse Laplace transform to $H(s)$. If the poles of the transfer function are known, then $H(s)$ can be expressed in partial fraction form:

$$H(s) = G + \frac{k_1}{s - p_1} + \frac{k_2}{s - p_2} + \cdots + \frac{k_n}{s - p_n}$$

$$= G + \sum_{i=1}^{n} \frac{k_i}{s - p_i}$$

(3.1.2)

where the coefficients k_i depend on the numerator of $H(s)$ as well as on the poles. The Laplace transform of a unit impulse centred on $t = 0$ is simply a constant:

$$\mathcal{L}\,\delta(t) = \int_{0-}^{\infty} \delta(t)e^{-st}\,\mathrm{d}t$$

$$= 1 \tag{3.1.3}$$

Also the Laplace transform of an exponential function of time is a first-order rational function of s:

$$\mathcal{L}e^{-\alpha t} = \int_{0-}^{\infty} e^{-(s+\alpha)t}\,\mathrm{d}t$$

$$= \frac{1}{s+\alpha} \tag{3.1.4}$$

Using these results allows the inverse Laplace transform of equation (3.1.2) to be determined:

$$h(t) = \mathcal{L}^{-1}H(s)$$

$$= G\,\delta(t) + \sum_{i=1}^{n} k_i e^{p_i t} \tag{3.1.5}$$

In the case of real poles, $p_i = \sigma_i$, the terms in the summation correspond to simple exponential decays:

$$e^{\sigma_i t}$$

Complex poles can be written in terms of their real and imaginary parts $p_i = \sigma_i + j\omega_i$. The corresponding terms in the impulse response are

$$e^{\sigma_i t}e^{j\omega_i t} = e^{\sigma_i t}\{\cos \omega_i t + j \sin \omega_i t\}$$

Of course the actual impulse response must be real; the imaginary parts of the terms from conjugate poles cancel out giving rise to terms in the impulse response of the form:

$$e^{\sigma_i t} \cos(\omega_i t + \phi_i)$$

It is clear that the real part of the poles determines the stability of a network. If the real part of any pole is positive, then the corresponding term in the impulse response will increase exponentially with time. All of the poles of a stable system must therefore lie on the left-hand side of the complex plane. The imaginary parts of the complex poles are the natural frequencies of the network.

Zeros of the transfer function do not determine the stability or natural frequencies of a network, but transfer functions which have no zeros in the right-half of the complex plane are minimum-phase transfer functions.

3.2   Factorization of a Polynomial

Consider the problem of factorizing a polynomial $p(x)$:

$$p(x) = a_0 + a_1 x + a_2 x^2 + \cdots + a_n x^n \tag{3.2.1}$$

where the coefficients $a_0 \ldots a_n$ are real. An alternative way of expressing $p(x)$ is as a product of n factors:

$$p(x) = g(x - z_1)(x - z_2) \ldots (x - z_n) \tag{3.2.2}$$

where the zeros $z_1 \ldots z_n$ are, in general, complex and are the roots of the equation

$$p(x) = 0 \tag{3.2.3}$$

Algebraic methods are available for finding the roots of polynomial equations of order 2, 3 and 4. Higher-order polynomial equations can only be solved by numerical methods.

Numerical procedures for evaluating the roots of polynomial equations do not attempt to find them all at once. Instead, one of the roots, say z_i, is located numerically. Then $(x - z_i)$, which must be a factor of the polynomial, is removed by a process of synthetic division. This is known as deflation, and reduces the order of the polynomial by one. By repeatedly finding a root and deflating the polynomial all of the zeros can be determined. The polynomial can then be written in the form of equation (3.2.2).

Since the zeros are, in general, complex, deflation will generate a polynomial with complex coefficients, even if the initial polynomial has real coefficients. When discussing procedures for finding the roots of polynomial equations, therefore, it will be assumed that the polynomial coefficients are complex.

All numerical root-finding methods are iterative. Starting with an initial approximation to the value of a root, the value is refined repeatedly by some process until it is known to a sufficient degree of accuracy. It is important that the roots be found in an ascending order of magnitude, otherwise, as will be demonstrated in section 3.6, the result of deflation will be of poor accuracy. Consequently the initial approximation should be an estimate of the smallest root of the polynomial.

Two root-finding methods will be discussed. The Newton–Raphson method converges rapidly towards a root, but may fail under certain circumstances. Svejgaard's method is more reliable, converging under all realistic circumstances, but is much less efficient than the Newton–Raphson method.

3.3 The Newton–Raphson Method

Although we are primarily concerned with finding the roots of polynomial equations, the Newton–Raphson method can be applied to any equation of the form:

$$f(x) = 0 \qquad (3.3.1)$$

Suppose that x_i is an initial estimate of a root of this equation. Then the function can be expanded as a Taylor series around this point:

$$f(x_i + h) = f(x_i) + hf'(x_i) + \frac{h^2}{2!}f''(x_i) + \cdots \qquad (3.3.2)$$

If $x_i + h$ is a root of the equation, then

$$f(x_i + h) = 0$$

Provided that the initial estimate x_i is close to the root, then h will be small, and the terms involving second and higher derivatives in equation (3.3.2) can be neglected. Thus

$$f(x_i) + hf'(x_i) = 0 \qquad (3.3.3)$$

In fact this equation will not give the distance h to the root exactly because of the approximation involved in neglecting the higher derivatives. It should, however, give a new point, $x_{i+1} = x_i + h$, which is closer to the root than the initial approximation. The Newton–Raphson iteration formula is therefore

$$x_{i+1} = x_i - f(x_i)/f'(x_i) \qquad (3.3.4)$$

This formula is applied iteratively until a result of sufficient accuracy is obtained. Convergence of the Newton–Raphson method is particularly rapid. Close to the root of a well-behaved function the number of correct digits doubles at each iteration.

A geometrical interpretation of the Newton–Raphson method is given in figure 3.1. This shows a typical function with an initial approximation x_0 to the root. The tangent to the function at the point x_0 is extrapolated back to the x axis, intersecting it at a new point x_1. Thus

$$\frac{\mathrm{d}f}{\mathrm{d}x} = \frac{f(x_0)}{x_0 - x_1}$$

which can be re-written in the form:

$$x_1 = x_0 - f(x_0)/f'(x_0)$$

This is just the Newton–Raphson iteration formula of equation (3.3.4). Figure 3.1 also confirms the rapid convergence of the Newton–Raphson method; after only three iterations the root has been located to an accuracy of better than 1%.

Figure 3.1
The Newton–
Raphson
method

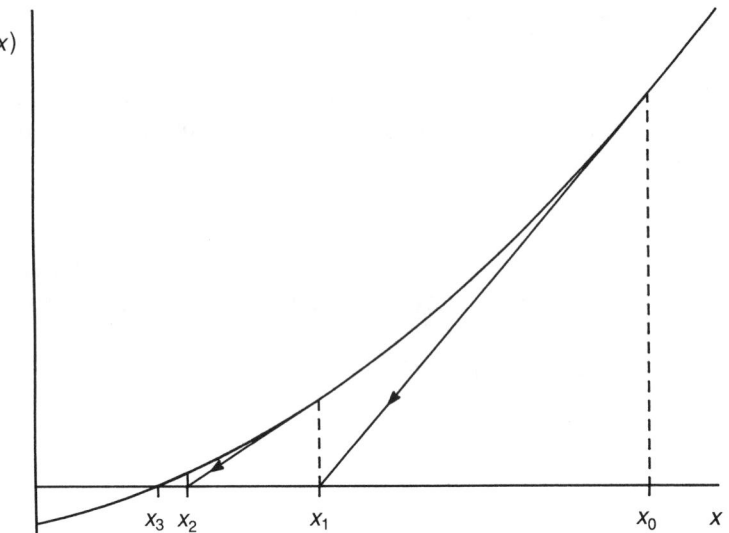

When the Newton–Raphson method is used to determine the roots of a polynomial the variable x, the function f and its derivative are all complex. This in no way affects the principle of the method but it does complicate root-finding programs. For the reasons discussed in the previous chapter, operations on complex variables will be written out explicitly in terms of their real and imaginary parts.

A new data type, *complexpoly*, is required to store the complex coefficients of the polynomials:

```
TYPE
    index = [0..maxsize];
    complex = RECORD re, im: REAL END;
    complexpoly = ARRAY index OF complex;
```

Evaluation of the polynomial can most efficiently be performed by the Horner scheme described in the previous chapter. In this case, however, the variable is complex rather than simply imaginary. A suitable procedure for performing the polynomial evaluations is shown below.

```
PROCEDURE evf(a: complexpoly; n: CARDINAL; x: complex;
            VAR f: complex);
VAR
    j: CARDINAL;
    w: REAL;
```

30

```
BEGIN
    f := a[n];
    FOR j := n-1 TO 0 BY -1 DO
        w := a[j].re+f.re*x.re-f.im*x.im;
        f.im := a[j].im+f.re*x.im+f.im*x.re;
        f.re := w
    END;
END evf;
```

In the case of polynomial functions the derivative can be obtained analytically:

$$\frac{d}{dx} \sum_{i=0}^{n} a_i x^i = \sum_{i=1}^{n} i a_i x^{i-1} \qquad (3.3.5)$$

A Horner evaluation scheme can therefore be used for the derivative:

```
PROCEDURE evd(a: complexpoly; n: CARDINAL; x: complex;
              VAR d: complex);
VAR
    j: CARDINAL;
    w: REAL;
BEGIN
    d.re:= FLOAT(n)*a[n].re; d.im := FLOAT(n)*a[n].im;
    IF n > 1 THEN
        FOR j := n-1 TO 1 BY -1 DO
            w := FLOAT(j)*a[j].re+d.re*x.re-d.im*x.im;
            d.im := FLOAT(j)*a[j].im+d.re*x.im+d.im*x.re;
            d.re := w
        END
    END
END evd;
```

An initial estimate of the magnitude of the smallest root is provided by the procedure *smallroot*. If a real starting value of x is used, and the polynomial coefficients are real, then the Newton–Raphson formula will generate a real result. Complex roots are therefore inaccessible from a real starting value, and a complex starting value of appropriate magnitude should be used.

Deciding when to terminate the Newton–Raphson iterations is by no means straightforward. The obvious criterion is the size of the correction term Δx relative to x, where

$$\Delta x = -f(x)/f'(x) \qquad (3.3.6)$$

Unfortunately, if a high degree of accuracy is demanded, then rounding errors in the calculation of Δx may prevent the condition for termination from ever being satisfied.

Close to a root the near-zero value of $f(x)$ results from the subtraction of nearly equal quantities, and the rounding error is related to the largest term in the polynomial. The corresponding rounding error in Δx depends on the value of $f'(x)$. Provided that there is substantially no cancellation between terms of the derivative polynomial then the rounding error in Δx has the same order of magnitude as the floating-point precision of x.

However, if there is another root close to x then there will be cancellation of derivative terms. Under these conditions the rounding error in Δx may be much larger than the precision of x. It is therefore necessary to choose an accuracy level that is well below the floating-point precision.

With a typical floating-point precision of 10^{-14}, a suitable condition for terminating the iterations might be

$$|\Delta x| < 10^{-9}|x| \qquad (3.3.7)$$

This criterion is likely to work in most cases but it must be accepted that it will, on occasion, fail.

A procedure for finding the roots of a polynomial by the Newton–Raphson method is given below:

```
PROCEDURE newton(VAR a: complexpoly; n: CARDINAL);

VAR
     m: CARDINAL;

     x, dx, fc, dc: complex;

     z: REAL;

BEGIN
     FOR m := n TO 1 BY -1 DO
          x.re := smallroot(a, m); x.im := x.re;
          IF x.re <> 0.0 THEN
               REPEAT
                    evf(a, m, x, fc);
                    evd(a, m, x, dc);
                    z := dc.re*dc.re+dc.im*dc.im;
                    dx.re := (fc.re*dc.re+fc.im*dc.im)/z;
                    dx.im := (fc.im*dc.re-fc.re*dc.im)/z;
                    x.re := x.re-dx.re; x.im := x.im-dx.im;
```

```
        UNTIL
            1.0E9*(ABS(dx.re)+ABS(dx.im)) < ABS(x.re)+ABS(x.im)
    END;
    deflate(a, m, x)
  END
END newton;
```

This procedure was used to find the roots of the polynomial

$$f(x) = x^4 - 10 \cdot 1 x^3 + 2 \cdot 0 x^2 - 10 \cdot 1 x + 1 \qquad (3.3.8)$$

During evaluation of the first zero, the following results were obtained on successive iterations:

$$9 \cdot 900990099010E - 002 \quad + \quad j \ 9 \cdot 900990099010E - 002$$
$$9 \cdot 901148280037E - 002 \quad - \quad j \ 1 \cdot 939659270894E - 003$$
$$9 \cdot 999974940521E - 002 \quad + \quad j \ 3 \cdot 751265632559E - 007$$
$$9 \cdot 999999999999E - 002 \quad - \quad j \ 1 \cdot 823871359041E - 014$$
$$1 \cdot 000000000000E - 001 \quad + \quad j \ 2 \cdot 839899258796E - 029$$

The very rapid convergence of the Newton–Raphson algorithm is clearly demonstrated. After a total of 34 polynomial evaluations, the four zeros of the polynomial were obtained:

$$1 \cdot 000000000000E - 001 \quad + \quad j \ 2 \cdot 839899258796E - 029$$
$$1 \cdot 382883313951E - 018 \quad + \quad j \ 1 \cdot 000000000000E + 000$$
$$7 \cdot 063663811252E - 018 \quad - \quad j \ 1 \cdot 000000000000E + 000$$
$$1 \cdot 000000000000E + 001 \quad - \quad j \ 1 \cdot 110223024625E - 016$$

The exact zeros are $0 \cdot 1$, j, $-j$ and 10.

Unfortunately, although the Newton–Raphson method works very well in most cases, there is no guarantee that it will always converge. Figure 3.2 shows what appears to be a well-behaved function, and yet, starting from the point x_0, the Newton–Raphson iterations are actually divergent.

Another condition under which the Newton–Raphson method may fail is where two or more of the zeros are coincident. Consider a polynomial function with m coincident zeros at z_0:

$$f(x) = g(x - z_0)^m (x - z_1) \dots (x - z_n)$$
$$= (x - z_0)^m f_1(x) \qquad (3.3.9)$$

Close to z_0 the function $f(x)$ tends to zero, as also does its derivative:

$$\frac{\mathrm{d}f}{\mathrm{d}x} = m(x - z_0)^{m-1} f_1(x) + (x - z_0)^m \frac{\mathrm{d}f_1}{\mathrm{d}x}$$

$$= (x - z_0)^{m-1} \left\{ m f_1(x) + (x - z_0) \frac{\mathrm{d}f_1}{\mathrm{d}x} \right\} \qquad (3.3.10)$$

33

Figure 3.2
Divergence of
the Newton–
Raphson
method

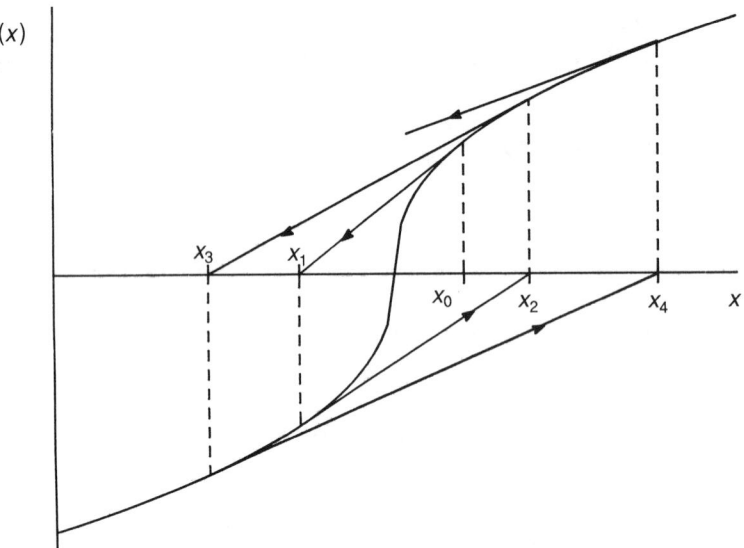

Both $f(x)$ and its derivative obtain their near-zero values from the subtraction of nearly equal terms and will therefore be of poor accuracy.

Close to coincident roots the convergence of the Newton–Raphson method will be unsatisfactory because of the poor accuracy of the correction term. There is also the danger that the derivative will be evaluated as zero, resulting in a divide-by-zero error when the correction term is calculated. For these reasons, the Newton–Raphson method for finding the roots of a polynomial equation cannot be considered to be of general application. It is, however, useful in specific cases where its rapid convergence makes it much more efficient than alternative, more general methods.

3.4 Svejgaard's Method

Many attempts have been made to devise a universal method for finding the roots of a polynomial equation. Most of these have been excessively complicated, catering for numerous special cases, and still failing to work under some conditions.

Svejgaard's method is an attempt to devise a root-finding algorithm founded on one simple general principle with no necessity to consider special cases. Although Svejgaard's method is far less efficient than the Newton–Raphson method, it is capable of dealing with multiple coincident roots, and fails only under the most extreme conditions. It is based on the fact that for a regular function $f(x)$ (any polynomial is regular), the minima of $|f(x)|$ are exactly the roots of the equation $f(x) = 0$.

A search procedure based on the principle of the rotating cross is used to locate the minima of $|f(x)|$. The modulus is evaluated at a starting point x, and at the four points $x + eh$, where $e^4 = 1$. These four points form a cross centred on x. If the moduli evaluated at the points of the cross are all greater than the modulus at the central point, then the minimum lies close to x; h is reduced in magnitude, rotated (by multiplying by $0 \cdot 4 + j0 \cdot 3$), and the process is repeated. On the other hand, if the modulus at one of the points of the cross is less than that at the central point then this replaces x; h is increased in magnitude (by multiplying by $1 \cdot 5$) and the process is repeated. Gradually the search converges on a minimum of $|f(x)|$, and is terminated when the size $|h|$ of the cross is sufficiently small relative to $|x|$.

A convenient criterion for terminating the iterations, which automatically adjusts to the floating-point precision, is the test of equality:

$$|x| = |x| + |h| \tag{3.4.1}$$

When the equality holds, the value of x has been found to the full precision of the floating-point representation. Compilers which perform a high level of optimization may however cancel $|x|$ from both sides of this test, leaving

$$0 = |h|$$

This is of course quite unsuitable, and leads to a large number of unnecessary iterations.

A small amount of computation can be saved by minimizing the square modulus of $f(x)$ rather than the modulus itself.

```
PROCEDURE modf(a: complexpoly; n: CARDINAL; x: complex): REAL;
VAR
    fc: complex;
BEGIN
    evf(a, n, x, fc);
    RETURN fc.re*fc.re+fc.im*fc.im
END modf;
```

A suitable starting point for the search is $x = 0$, with the size of the cross h equal to the magnitude of the smallest root. The procedure shown below finds the roots of a polynomial by Svejgaard's method:

```
PROCEDURE svejgaard(VAR a: complexpoly; n: CARDINAL);
VAR
    i, m: CARDINAL;
    dx, x, xx, x0: complex;
    t, f0, f1: REAL;
    centre: BOOLEAN;
```

```
BEGIN
    FOR m := n TO 1 BY -1 DO
        dx.re := smallroot(a, m); dx.im := 0.0;
        x.re := 0.0; x.im := 0.0;
        REPEAT
            f0 := modf(a, m, x);
            centre := TRUE;
            FOR i := 1 TO 4 DO
                xx.re := x.re+dx.re; xx.im := x.im+dx.im;
                f1 := modf(a, m, xx);
                IF f1 < f0 THEN
                    f0 := f1; centre := FALSE; x0 := xx
                END;
                t := -dx.im; dx.im := dx.re; dx.re := t
            END;
            IF centre THEN
                t := 0.4*dx.re-0.3*dx.im;
                dx.im := 0.4*dx.im+0.3*dx.re;
                dx.re := t
            ELSE
                dx.re := 1.5*dx.re;
                dx.im := 1.5*dx.im;
                x := x0
            END;
            f0 := ABS(x.re)+ABS(x.im);
            f1 := ABS(dx.re)+ABS(dx.im);
        UNTIL f0 = f0+f1;
        deflate(a, m, x)
    END
END svejgaard;
```

When this procedure was used to determine the roots of equation (3.3.8) the
following result was obtained after 2320 polynomial evaluations:

$$1 \cdot 000000000000E-001 \quad - \quad j\ 3 \cdot 717729459862E-023$$
$$0 \cdot 000000000000E+000 \quad + \quad j\ 1 \cdot 000000000000E+000$$
$$2 \cdot 853160812953E-022 \quad - \quad j\ 1 \cdot 000000000000E+000$$
$$1 \cdot 000000000000E+001 \quad + \quad j\ 2 \cdot 693321849137E-020$$

By comparison the Newton–Raphson method required only 34 evaluations. However, the procedure based on Svejgaard's method coped successfully with a polynomial which had eight coincident roots, whereas the Newton–Raphson procedure failed with a polynomial which had only three coincident roots.

3.5 Estimation of the Smallest Root

All iterative root-finding methods must be provided with a starting point. In this section a method for estimating the magnitude of the smallest root will be described.

Suppose that the roots of a polynomial equation are $z_1, z_2, \ldots z_n$ and that they are ordered according to

$$|z_1| \leqslant |z_2| \leqslant |z_3| \leqslant \cdots \leqslant |z_n| \tag{3.5.1}$$

The corresponding equation will be

$$(x - z_1)(x - z_2) \ldots (x - z_n) = 0 \tag{3.5.2}$$

Dividing through by the product of the roots gives

$$(1 - x/z_1)(1 - x/z_2) \ldots (1 - x/z_n) = 0 \tag{3.5.3}$$

Multiplying out the factors in this equation, and comparing coefficients with the polynomial equation:

$$\sum_{i=0}^{n} a_i x^i = 0 \tag{3.5.4}$$

leads to the following approximate results:

$$\left| \frac{a_1}{a_0} \right| \approx \frac{1}{|z_1|}$$

$$\left| \frac{a_2}{a_0} \right| \approx \frac{1}{|z_1||z_2|}$$

$$\left| \frac{a_3}{a_0} \right| \approx \frac{1}{|z_1||z_2||z_3|} \tag{3.5.5}$$

$$\vdots$$

$$\left| \frac{a_n}{a_0} \right| \approx \frac{1}{|z_1||z_2||z_3| \ldots |z_n|}$$

The approximate magnitude of the smallest root z_1 should therefore be given by

$$|z_1| = \left| \frac{a_1}{a_0} \right|^{-1} \tag{3.5.6}$$

37

However, if z_1 and z_2 are of similar magnitude then it is possible for the coefficient a_1 to approach zero by cancellation of terms. In that case the magnitude of z_1 can be estimated from the coefficient a_2:

$$|z_1| \approx \{|z_1||z_2|\}^{1/2} \approx \left|\frac{a_2}{a_0}\right|^{-1/2} \tag{3.5.7}$$

If, in turn, a_2 is nearly zero because of cancellation, then the coefficient a_3 can be used, and so on.

A general formula for the magnitude of the smallest root is therefore

$$|z_1| \approx \min_{i=1}^{n} \left|\frac{a_i}{a_0}\right|^{-1/i} \tag{3.5.8}$$

Modula-2 does not provide for raising variables to fractional powers, but the operation can be performed by using the logarithm and exponential functions as shown below:

```
PROCEDURE smallroot(a: complexpoly; n: CARDINAL): REAL;
VAR
    i: CARDINAL;
    t1, t2, z: REAL;
BEGIN
    IF (a[0].re = 0.0) AND (a[0].im = 0.0) THEN RETURN 0.0 END;
    z := infinity;
    FOR i := 1 TO n DO
        t1 := (ABS(a[i].re)+ABS(a[i].im))/
            (ABS(a[0].re)+ABS(a[0].im));
        IF t1 > 0.0 THEN
            t2 := exp(-ln(t1)/FLOAT(i));
            IF z > t2 THEN z := t2 END
        END
    END;
    RETURN z;
END smallroot;
```

The estimates produced by *smallroot* during the process of finding the roots of equation (3.3.8) are

$$9 \cdot 900990099010\text{E} - 002$$
$$1 \cdot 000000000000\text{E} + 000$$
$$9 \cdot 090909090909\text{E} - 001$$
$$1 \cdot 000000000000\text{E} + 001$$

These values are good approximations to the actual magnitudes of the roots of the equation, which are $0 \cdot 1$, j, $-j$ and 10.

3.6 Polynomial Deflation

If z_1 is a root of a polynomial equation, then $(x - z_1)$ must be a factor of the polynomial, and can therefore be removed by division. It is essential that the correct procedure is used, otherwise a serious loss of accuracy can result.

Consider the polynomial shown below which has roots at $0 \cdot 001$ and 1000:

$$x^2 - 1000 \cdot 001 x + 1 \cdot 0 \qquad (3.6.1)$$

The factor $(x - 0 \cdot 001)$ corresponding to the smallest root can be divided out of this polynomial;

$$
\begin{array}{r|l}
 & x - 1000 \cdot 0 \\
\hline
x - 0 \cdot 001 & x^2 - 1000 \cdot 001 x + 1 \cdot 0 \\
 & x^2 - 0 \cdot 001 x \\
\hline
 & -1000 \cdot 0 \quad x + 1 \cdot 0 \\
 & -1000 \cdot 0 \quad x + 1 \cdot 0 \\
\hline
\end{array}
$$

As expected, the polynomial left after deflation is $x - 1000$. None of the coefficients in the result were derived from the subtraction of nearly equal quantities.

However, if the factor corresponding to the largest root is divided out of the polynomial the situation is quite different:

$$
\begin{array}{r|l}
 & x - 0 \cdot 001 \\
\hline
x - 1000 \cdot 0 & x^2 - 1000 \cdot 001 x + 1 \cdot 0 \\
 & x^2 - 1000 \cdot 0 \quad x \\
\hline
 & -0 \cdot 001 x + 1 \cdot 0 \\
 & -0 \cdot 001 x + 1 \cdot 0 \\
\hline
\end{array}
$$

In this case the coefficient $0 \cdot 001$ in the result is derived from the subtraction of $1000 \cdot 0$ from $1000 \cdot 001$, and the absolute accuracy of the result is related to the rounding error in 1000; this is 10^6 times the floating-point precision of $0 \cdot 001$. Consequently six digits of precision have been lost in this deflation process.

These examples illustrate the general principle that if deflation is to be performed with minimal loss of accuracy, then the smallest root of the polynomial equation should be evaluated first.

A procedure for dividing a complex root x from a complex polynomial a is given below:

```
PROCEDURE deflate(VAR a: complexpoly; n: CARDINAL; x: complex);
VAR
    j: CARDINAL;
    s, t, u, v: REAL;
BEGIN
    s := 0.0; t := 0.0;
    FOR j := n TO 0 BY -1 DO
        WITH a[j] DO
            u := re+x.re*s-x.im*t;
            v := im+x.re*t+x.im*s;
            re := u; s := u;
            im := v; t := v
        END
    END;
    FOR j := 0 TO n-1 DO a[j] := a[j+1] END;
    a[n] := x
END deflate;
```

An alternative deflation procedure can be devised which removes a factor corresponding to a large root from a polynomial without serious loss of accuracy; this procedure fails to remove a small root satisfactorily, however.

3.7 Summary

Linear networks with no distributed elements have transfer functions that are the ratio of two polynomials in s. Zeros of the numerator polynomial are also transfer function zeros; zeros of the denominator polynomial are transfer function poles. The poles, and to a lesser extent the zeros, of a transfer function determine its transient response. In particular, a transfer function with one or more positive real-part poles is unstable.

Polynomial equations of order 5 and higher must be solved by numerical methods. Zeros are normally found one at a time and removed from the polynomial by deflation. This process is repeated until all of the zeros have been determined. All numerical zero-finding procedures are iterative. Starting from some initial estimate of a zero, the value is refined at each stage until a sufficient accuracy is attained. In general, the zeros of a polynomial are complex; following deflation, therefore, the polynomial coefficients will also become complex.

The Newton–Raphson method for finding polynomial zeros is very efficient, particularly if it is provided with a good starting point. It can, however, fail to converge under some circumstances. If two or more of the polynomial zeros are coincident, then close to these zeros both the polynomial and its derivative become very small and of poor accuracy. As a result the Newton–Raphson iteration may be divergent.

Svejgaard's method is based on the fact that the minima of the modulus of a polynomial are the zeros of the polynomial. A two-dimensional search for the minima is conducted in the complex plane using the principle of the rotating cross. Svejgaard's method is very reliable, and can deal with coincident zeros, but is much less efficient than the Newton–Raphson method.

Deflation is the process by which zeros are removed from polynomials. It consists effectively of synthetic division and reduces the order of a polynomial by one. The most accurate results are obtained when deflation is performed using the smallest zero. At each stage, therefore, the starting point for either the Newton–Raphson or Svejgaard's algorithm should be an estimate of the smallest zero. This can easily be determined from the polynomial coefficients.

3.8 Problems

1 Determine which of the following 4th-order transfer functions represent stable systems:

$$H(s) = \frac{1}{4 + 10s + 10s^2 + 5s^3 + s^4}$$

$$H(s) = \frac{1}{12 - 2s - 2s^2 + 3s^3 + s^4}$$

$$H(s) = \frac{1}{15 + 26s + 16s^2 + 6s^3 + s^4}$$

$$H(s) = \frac{1}{8 + 10s + 3s^2 + 2s^3 + s^4}$$

2 The polynomial given below has four coincident zeros at $x = 1$:

$$1 - 4x + 6x^2 - 4x^3 + x^4$$

Use both the Newton–Raphson method and Svejgaard's method to find the zeros of this polynomial. Compare the accuracy and efficiency of the two methods.

3 A polynomial equation which resists root-finding by both the Newton–Raphson method and Svejgaard's method is given below:

$$f(x) = x^n + 1 = 0$$

where n is a large integer. The roots of this equation can, of course, be determined analytically and lie on the unit circle of the complex plane.

Determine the maximum value of n for which each of the root-finding methods works successfully, and investigate the reasons for failure with higher values of n.

4   Figure 3.3 shows the block diagram of a simple feedback control system. The overall transfer function $H(s)$ of this control system is given by

$$H(s) = \frac{kG(s)}{1 + kF(s)G(s)}$$

Given that

$$F(s)G(s) = \frac{1}{(s+1)(s+3)(s+4)}$$

plot on the complex plane the positions of the transfer function poles as k varies from $0 \cdot 1$ to 1000.

Figure 3.3
A feedback
control system

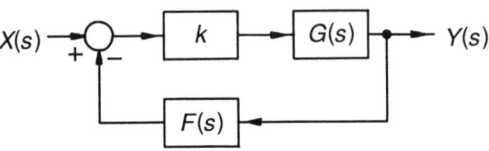

5   The square root of a number k can be obtained by solving the polynomial equation:

$$f(x) = x^2 - k = 0$$

Using the Newton–Raphson method, derive an iterative formula for evaluating the square root of k. The formula should include only the operations of addition, subtraction, multiplication and division.

Calculate the square root of 3 to eight decimal digits of accuracy. Investigate the convergence of the formula starting from $x_0 = 2$, $x_0 = 20$, $x_0 = 2000$.

Suppose that the iterative formula is to be used to calculate the square roots of numbers between 10^{-100} and 10^{100}. How can a suitable starting point be obtained?

4 Numerical Analysis of Linear Networks

4.1 Introduction

Numerical network analysis is concerned with obtaining the network functions of linear networks as complex numbers. Network functions fall into two classes: input functions and transfer functions. Input functions are input impedances or input admittances; transfer functions are voltage transfer ratios, current transfer ratios, transimpedances or trans-admittances. Consider the bridged-T filter network shown in figure 4.1.

This network can be analyzed to obtain the symbolic form of the voltage frequency response function:

$$H(j\omega) = \frac{1 + C_a(R_c + R_d)j\omega + R_cR_dC_aC_b(j\omega)^2}{1 + C_a(R_c + R_d)j\omega + C_bR_cj\omega + R_cR_dC_aC_b(j\omega)^2} \qquad (4.1.1)$$

When a network is analyzed manually it is usually obtained in symbolic form. Symbolic network functions can be reduced to numerical form by substitution of the component values. For the network shown in figure 4.1 the numerical form of the transfer function is given by

$$H(j\omega) = \frac{1 + (2 \times 10^{-3}j\omega) + 10^{-6}(j\omega)^2}{1 + (3 \times 10^{-3}j\omega) + 10^{-6}(j\omega)^2} \qquad (4.1.2)$$

Figure 4.1
A bridged-T filter

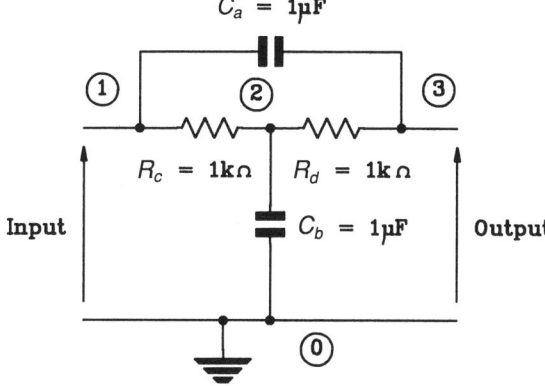

$C_a = 1\mu F$

① ② ③

$R_c = 1k\Omega$ $R_d = 1k\Omega$

Input $C_b = 1\mu F$ Output

⓪

43

This form of the network function contains less information than the symbolic form, and its dependence on the values of the individual components is no longer directly accessible.

Substitution of a value for ω into a numerical network function yields a complex number; in the case of a voltage transfer function this represents the gain and phase at the frequency ω. For example, at a frequency of $\omega = 2 \times 10^3$ rad/s the voltage transfer ratio of the bridged-T filter is given by

$$H(2 \times 10^3 \text{j}) = \frac{-3 + \text{j}4}{-3 + \text{j}6} = \frac{11 + \text{j}2}{15}$$

so that gain $= -2 \cdot 55$ dB and phase $= 10 \cdot 3°$.

Again, information has been lost as a result of this substitution process. In spite of the fact that a table giving the variation of gain and phase with frequency contains less information than the corresponding symbolic transfer function, it is still in this form that the network behaviour is often required to be known. For example, a table of gain against frequency could be used to determine whether a filter network met some frequency-domain specification.

Computer methods are available for determining network functions in symbolic form, and these will be discussed in the following chapter. They are, however, considerably more complex, and in some circumstances less efficient, than numerical methods.

Many of the methods used to analyze networks manually are difficult to adapt for implementation by computer. For example, when analyzing a network manually it may be possible to use Thévenin's theorem or a star–delta transformation to reduce the amount of algebraic manipulation. However, it is very difficult to program a computer to recognize when such simplification techniques should be employed. In practice, computer analysis is performed by methods which are sufficiently general to deal with networks of any complexity and which do not attempt to exploit particular network configurations.

The numerical network analysis method that will be described here is known as nodal analysis. Nodes are points in a network where components terminate and several components can terminate at a single node. The network shown in figure 4.1 contains four nodes, numbered 0 ... 3. By convention, the reference node (or ground node) relative to which the voltages on the other nodes are measured is assigned the number zero.

For the sake of generality it will be assumed that external current generators are applied to all of the nodes of the network. It is recognized, of course, that in most cases all of the external generators, except the one connected to the input node, will have zero values. Kirchhoff's current law is applied to each of the nodes except the reference node, yielding a set of simultaneous linear equations in which the unknown values are the nodal voltages. These nodal equations are usually expressed in matrix-vector form. If **V** is a column vector of nodal voltages measured with respect to the reference node, and **I** is a column vector of external current generators, then

$$\mathbf{Y}\,\mathbf{V} = \mathbf{I} \qquad\qquad (4.1.3)$$

Y is a square matrix whose elements have the dimension of admittance. It is known as the nodal admittance matrix, or sometimes simply as the Y-matrix. Solving these equations yields the nodal voltages from which the input impedance, transimpedance and voltage transfer ratio can be deduced.

The first step in performing a nodal analysis of a network is the determination of the Y-matrix. This can be done formally with the aid of network graphs. It will be assumed initially that the network contains only two-terminal passive components. Later on the method will be generalized to include controlled sources and infinite-gain operational amplifiers.

4.2   The Graphs of Passive Networks

An electrical network consists of interconnected components such as resistors, capacitors, inductors and controlled sources. Many of the properties of a network are determined by its topology and are independent of the component values. A convenient representation of network topology is the network graph (or graphs).

A graph consists of nodes (which correspond directly with the nodes of the network) and branches which are line segments connecting pairs of nodes. Networks are normally represented by directed graphs (sometimes called digraphs) in which each branch is assigned a direction or orientation. An example of a directed graph is shown in figure 4.2. The nodes are numbered 0 ... 3 and the branches are labelled $a ... e$.

The exact way in which a graph is drawn is unimportant, so that the graphs shown in figure 4.3 are to be regarded as identical with that in figure 4.2.

In the case of a passive transformerless network the network graph is obtained simply by replacing each of the two-terminal components by a branch. The choice of branch direction is entirely arbitrary. Figure 4.4 shows the network graph of the bridged-T filter network.

Figure 4.2
A network
graph

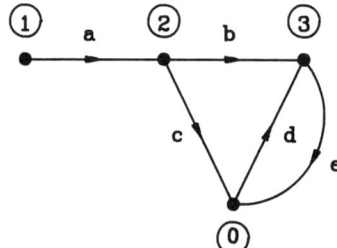

45

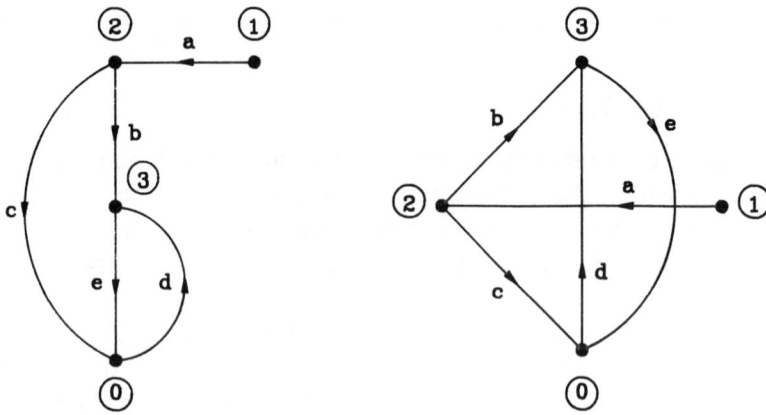

Figure 4.3
Alternative
network graphs

Figure 4.4
Network graph
of the bridged-
T filter

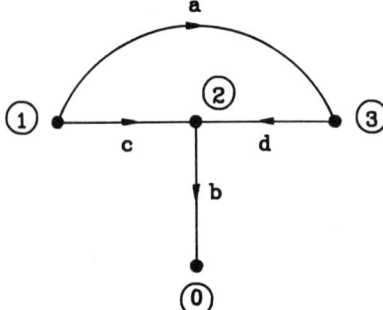

Associated with each passive component y_r of the network will be a voltage v_r and a current i_r as shown in figure 4.5. The convention used is that the current flows in the direction of the corresponding branch of the network graph. Ohm's law provides a relationship between the individual component voltages and currents:

$$i_r = y_r v_r \qquad (4.2.1)$$

In the case of the bridged-T filter there are four such equations:

$$\begin{aligned} i_a &= y_a v_a \\ i_b &= y_b v_b \\ i_c &= y_c v_c \\ i_d &= y_d v_d \end{aligned} \qquad (4.2.2)$$

Figure 4.5
A passive
component

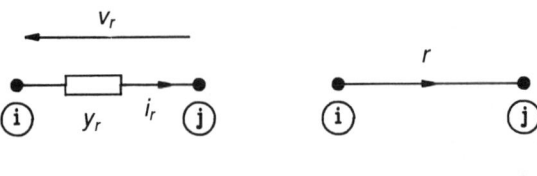

Network　　　　　　　**Graph**

46

These equations can be written in matrix form:

$$
\begin{bmatrix} i_a \\ i_b \\ i_c \\ i_d \end{bmatrix} = \begin{bmatrix} y_a & 0 & 0 & 0 \\ 0 & y_b & 0 & 0 \\ 0 & 0 & y_c & 0 \\ 0 & 0 & 0 & y_d \end{bmatrix} \begin{bmatrix} v_a \\ v_b \\ v_c \\ v_d \end{bmatrix}
\tag{4.2.3}
$$

or

$$
\mathbf{i} = \mathbf{C}\,\mathbf{v}
\tag{4.2.4}
$$

The diagonal matrix $\mathbf{C}$ is known as the branch admittance matrix. Its elements have dimensions of admittance, and its size is $m \times m$, where m is the number of passive components. $\mathbf{i}$ and $\mathbf{v}$ are the branch current and branch voltage vectors respectively; they are both of size $m \times 1$.

It will be assumed that independent external currents are applied to each node of the network. Kirchhoff's current law requires that the current flowing into any node must equal the current flowing out, and for the bridged-T filter this leads to the following set of equations:

$$
\begin{aligned}
-i_b &&&= I_0 \\
i_a && +i_c &&&= I_1 \\
&& i_b & -i_c & -i_d &= I_2 \\
-i_a &&&& +i_d &= I_3
\end{aligned}
\tag{4.2.5}
$$

where $I_0 \ldots I_3$ are the external currents. These equations are not independent, and any one of the equations can be derived from the others by taking linear combinations. This can be seen by summing the left-hand sides of the four equations. The result must be zero because each component current appears twice, once with a positive sign and once with a negative sign. (The sum of the right-hand sides $I_0 + I_1 + I_2 + I_3$ is the net current flowing into the network and must also be zero.) One of the equations can therefore be ignored without any loss of information. Normally it is the equation for the reference node that is discarded, and the remaining equations can be written in matrix form:

$$
\begin{bmatrix} 1 & 0 & 1 & 0 \\ 0 & 1 & -1 & -1 \\ -1 & 0 & 0 & 1 \end{bmatrix} \begin{bmatrix} i_a \\ i_b \\ i_c \\ i_d \end{bmatrix} = \begin{bmatrix} I_1 \\ I_2 \\ I_3 \end{bmatrix}
\tag{4.2.6}
$$

or

$$
\mathbf{A}\,\mathbf{i} = \mathbf{I}
\tag{4.2.7}
$$

The matrix $\mathbf{A}$ is known as the incidence matrix (or, strictly, the reduced incidence matrix) and contains all the information concerning the network topology. It is of size $n \times m$ where n is the number of nodes excluding the reference node. The elements A_{ij} accord with the following rules:

$$A_{ij} = \quad 0 \quad \text{if branch } j \text{ is not connected to node } i$$
$$A_{ij} = +1 \quad \text{if branch } j \text{ is directed away from node } i \qquad (4.2.8)$$
$$A_{ij} = -1 \quad \text{if branch } j \text{ is directed towards node } i$$

A network contains two sorts of voltages, the nodal voltages V_i and the branch voltages v_j. In the case of the bridged-T filter they are related by the following set of equations:

$$V_1 - V_3 = v_a$$
$$V_2 - V_0 = v_b$$
$$V_1 - V_2 = v_c \qquad (4.2.9)$$
$$V_3 - V_2 = v_d$$

Now the voltage V_0 on the reference node is by definition zero, and the equations can therefore be written in matrix form:

$$\begin{bmatrix} 1 & 0 & -1 \\ 0 & 1 & 0 \\ 1 & -1 & 0 \\ 0 & -1 & 1 \end{bmatrix} \begin{bmatrix} V_1 \\ V_2 \\ V_3 \end{bmatrix} = \begin{bmatrix} v_a \\ v_b \\ v_c \\ v_d \end{bmatrix} \qquad (4.2.10)$$

or

$$\mathbf{B} \, \mathbf{V} = \mathbf{v} \qquad (4.2.11)$$

The matrix $\mathbf{B}$ is of size $m \times n$ and its elements B_{ij} accord with the following rules:

$$B_{ij} = \quad 0 \quad \text{if branch } i \text{ is not connected to node } j$$
$$B_{ij} = +1 \quad \text{if branch } i \text{ is directed away from node } j \qquad (4.2.12)$$
$$B_{ij} = -1 \quad \text{if branch } i \text{ is directed towards node } j$$

From these rules, and those given in (4.2.8), it is clear that

$$A_{ij} = B_{ji} \qquad (4.2.13)$$

Matrix $\mathbf{B}$ is therefore simply the transpose of the incidence matrix $\mathbf{A}$, and equation (4.2.11) becomes:

$$\mathbf{A}^T \mathbf{V} = \mathbf{v} \qquad (4.2.14)$$

Finally, equations (4.2.4), (4.2.7) and (4.2.14) can be combined to give

$$\mathbf{A} \, \mathbf{C} \, \mathbf{A}^T \mathbf{V} = \mathbf{I} \qquad (4.2.15)$$

or

$$\mathbf{Y} \, \mathbf{V} = \mathbf{I} \qquad (4.2.16)$$

where $\mathbf{Y} = \mathbf{A} \, \mathbf{C} \, \mathbf{A}^T \qquad (4.2.17)$

$\mathbf{Y}$ is of size $n \times n$ and is known as the nodal admittance matrix, or simply as the Y-matrix.

It is interesting to consider what contribution each component makes to the Y-matrix. Consider the case where the rth admittance is connected between node i and the reference node 0 as shown in figure 4.6a. The Y-matrix for this component alone is given by

$$
\mathbf{Y} = i \begin{matrix} & r \\ \begin{bmatrix} 0 & 0 & 0 \\ 0 & 1 & 0 \\ 0 & 0 & 0 \end{bmatrix} \end{matrix} \begin{bmatrix} 0 & & \\ & y_r & \\ & & 0 \end{bmatrix} \begin{matrix} & i \\ \begin{bmatrix} 0 & 0 & 0 \\ 0 & 1 & 0 \\ 0 & 0 & 0 \end{bmatrix} \end{matrix} \begin{matrix} \\ \\ r \end{matrix}
$$

$$
= i \begin{matrix} & i & \\ \begin{bmatrix} 0 & 0 & 0 \\ 0 & y_r & 0 \\ 0 & 0 & 0 \end{bmatrix} \end{matrix} \tag{4.2.18}
$$

In other words, the admittance contributes only to the diagonal element Y_{ii}.

Now consider the case of an admittance connected between two nodes, i and j, neither of which is the reference node as shown in figure 4.6b. The Y-matrix for this component is given by

$$
\mathbf{Y} = \begin{matrix} i \\ j \end{matrix} \begin{matrix} & r & \\ \begin{bmatrix} 0 & 0 & 0 \\ 0 & 1 & 0 \\ 0 & -1 & 0 \\ 0 & 0 & 0 \end{bmatrix} \end{matrix} \begin{bmatrix} 0 & & \\ & y_r & \\ & & 0 \end{bmatrix} \begin{matrix} & i & j & \\ \begin{bmatrix} 0 & 0 & 0 & 0 \\ 0 & 1 & -1 & 0 \\ 0 & 0 & 0 & 0 \end{bmatrix} \end{matrix} \begin{matrix} \\ \\ r \end{matrix}
$$

$$
= \begin{matrix} i \\ j \end{matrix} \begin{matrix} & i & j \\ \begin{bmatrix} y_r & -y_r \\ -y_r & y_r \end{bmatrix} \end{matrix} \tag{4.2.19}
$$

Figure 4.6
Components in a passive network

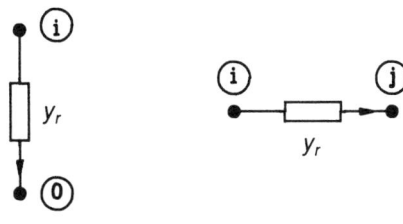

a

b

49

This admittance contributes to the diagonal elements Y_{ii} and Y_{jj}, and with a negative sign to the off-diagonal elements Y_{ij} and Y_{ji}.

Since the network is assumed to be linear, the complete Y-matrix is simply the sum of the contributions from the individual admittances. A general rule for setting up the Y-matrix for a passive network can be derived from equations (4.2.18) and (4.2.19):

Y_{ii} = sum of all admittances connected to node i $(i = 1 \dots n)$
Y_{ij} = *minus* sum of all admittances connected between nodes (4.2.20)
i and j $(i, j = 1 \dots n; \ i \neq j)$

It is therefore a fairly simple matter to program a computer to determine the Y-matrix from a network description, either by setting up the branch admittance and incidence matrices and performing the appropriate matrix multiplications, or by using the rules given in (4.2.20).

4.3 Solution of the Nodal Equations

Multiplication of both sides of equation (4.2.16) by the inverse of the Y-matrix gives the node voltages in terms of the external current generators:

$$\mathbf{V} = \mathbf{Y}^{-1} \, \mathbf{I} \qquad (4.3.1)$$

Usually we are interested in the voltage V_s generated at some node s which results from an input applied to the network at another node t. In this case all of the external current generators are set to zero except I_t. Row s of the matrix equation (4.3.1) can be written:

$$\begin{aligned} V_s &= [\mathbf{Y}^{-1}]_{s1} I_1 + \cdots + [\mathbf{Y}^{-1}]_{st} I_t + \cdots + [\mathbf{Y}^{-1}]_{sn} I_n \\ &= [\mathbf{Y}^{-1}]_{st} I_t \end{aligned} \qquad (4.3.2)$$

This result gives the transimpedance V_s / I_t of the network. By setting $s = t$ in equation (4.3.2), an expression for the input impedance is obtained:

$$V_t = [\mathbf{Y}^{-1}]_{tt} I_t \qquad (4.3.3)$$

Finally, combining equations (4.3.2) and (4.3.3) gives the voltage transfer ratio of the network:

$$\frac{V_s}{V_t} = \frac{[\mathbf{Y}^{-1}]_{st}}{[\mathbf{Y}^{-1}]_{tt}} = \frac{\Delta_{ts}}{\Delta_{tt}} \qquad (4.3.4)$$

where Δ_{ij} is the (i, j) cofactor of the Y-matrix.

Thus the voltage gain of a network can be calculated by numerical inversion of the Y-matrix. However, although equation (4.3.4) is an important result, and will be used in the next chapter on symbolic network analysis, it does not form the basis of an efficient method for calculating the gain. The reason for this is that numerical matrix inversion is computationally more complex than solving an equivalent number of simultaneous equations. A set of simultaneous equations can be solved using $n^3/3$ multiplications, whereas inversion of an $n \times n$ matrix requires n^3 multiplications.

It is therefore more efficient to solve the nodal equations (4.2.16) directly. Before this can be done, suitable numerical values must be assigned to the elements of the external current vector **I**. All of the elements of **I** are set to zero except I_t. Since the network is assumed to be linear, the value of I_t is not critical (it will cancel out of the final result) and a value of 1 A can be used. Solution of the equations gives a nodal voltage vector **V** whose elements represent the responses to a 1 A input at node t. From these voltages the following properties of the network can be calculated:

Transimpedance	$V_s/1$ A	
Input impedance	$V_t/1$ A	(4.3.5)
Voltage transfer ratio	V_s/V_t	

Almost all methods for solving simultaneous linear equations are based on the elementary principle of Gaussian elimination. The first equation is used to eliminate the first variable from the second and subsequent equations. Then the second equation is used to eliminate the second variable from the third and subsequent equations. This process is repeated until the nth equation contains only the nth variable. The first stage of the Gaussian elimination is now complete. In the second stage, the value of the nth variable is back-substituted into equations $1 \ldots (n-1)$. Since the $(n-1)$th equation now contains only the $(n-1)$th variable, it can in turn be back-substituted. When all of the variables have been back-substituted, the process is effectively complete, with each equation containing only a single variable.

Consider the set of n simultaneous equations:

$$
\begin{aligned}
A_{11}x_1 + A_{12}x_2 + A_{13}x_3 + \cdots + A_{1n}x_n &= b_1 \\
A_{21}x_1 + A_{22}x_2 + A_{23}x_3 + \cdots + A_{2n}x_n &= b_2 \\
A_{31}x_1 + A_{32}x_2 + A_{33}x_3 + \cdots + A_{3n}x_n &= b_3 \\
&\vdots \\
A_{n1}x_1 + A_{n2}x_2 + A_{n3}x_3 + \cdots + A_{nn}x_n &= b_n
\end{aligned}
\tag{4.3.6}
$$

or

$$
\mathbf{A} \, \mathbf{x} = \mathbf{b}
\tag{4.3.7}
$$

In order to eliminate the variable x_1 from equation j, equation 1 is multiplied by A_{j1}/A_{11}, and subtracted from equation j. After this operation has been performed for $j = 2$ to $j = n$ the following set of equations is obtained:

$$A'_{11}x_1 + A'_{12}x_2 + A'_{13}x_3 + \cdots + A'_{1n}x_n = b'_1$$
$$A'_{22}x_2 + A'_{23}x_3 + \cdots + A'_{2n}x_n = b'_2$$
$$A'_{32}x_2 + A'_{33}x_3 + \cdots + A'_{3n}x_n = b'_3 \qquad (4.3.8)$$
$$\vdots$$
$$A'_{n2}x_2 + A'_{n3}x_3 + \cdots + A'_{nn}x_n = b'_n$$

A similar procedure can now be used to remove variable x_2; equation 2 is multiplied by A_{j2}/A_{22} and subtracted from equation j. In general, if the variable x_i is being removed from equation j, then the kth coefficient on the left-hand side of this equation undergoes the transformation:

$$A_{jk} := A_{jk} - A_{ji}A_{ik}/A_{ii} \qquad (4.3.9)$$

On completion of this procedure for each of the variables $x_1 \ldots x_{n-1}$, the set of equations assumes a triangular form:

$$A''_{11}x_1 + A''_{12}x_2 + A''_{13}x_3 + \cdots + A''_{1n}x_n = b''_1$$
$$A''_{22}x_2 + A''_{23}x_3 + \cdots + A''_{2n}x_n = b''_2$$
$$A''_{33}x_3 + \cdots + A''_{3n}x_n = b''_3 \qquad (4.3.10)$$
$$\vdots$$
$$A''_{nn}x_n = b''_n$$

Only variable x_n can be evaluated directly from this set of equations:

$$x_n = b''_n/A''_{nn} \qquad (4.3.11)$$

However by back-substitution of x_n it can be removed from the other equations; variable x_{n-1} can then be evaluated directly. When variable x_i is back-substituted into equation j, the right-hand side undergoes a transformation:

$$b''_j := b''_j - b''_i A''_{ji}/A''_{ii} \qquad (4.3.12)$$

By repeating the back-substitution process for x_n to x_2, all of the variables can be evaluated.

Unfortunately the Gaussian elimination procedure described above can fail under certain circumstances. In particular, if the value of A_{ii} in assignment (4.3.9) happens to be zero, then a divide-by-zero error will result. A_{ii} is known as the pivot and, even if its value is not zero, a severe loss of accuracy can result if its magnitude is very small. To see why this should be the case, consider the following set of two simultaneous equations:

$$0 \cdot 001 x_1 - 2x_2 = 4 \cdot 001 \qquad (4.3.13)$$
$$3x_1 + x_2 = 1$$

It can easily be confirmed that the solution of these equations is

$$x_1 = 1 \tag{4.3.14}$$
$$x_2 = -2$$

Now suppose that the equations are to be solved by a computer using a floating-point representation that is accurate only to three decimal digits. The equations will be stored in the computer as

$$0 \cdot 001 x_1 - 2 x_2 = 4 \tag{4.3.15}$$
$$3 x_1 + \quad x_2 = 1$$

Variable x_1 is eliminated by multiplying the first equation by $3/0 \cdot 001$ and subtracting from the second equation; the results are truncated to three decimal digits:

$$0 \cdot 001 x_1 - 2 x_2 = 4 \tag{4.3.16}$$
$$6000 x_2 = -12000$$

A correct value of $x_2 = -2$ is obtained from the second equation. However, when this is back-substituted into the first equation the result is $x_1 = 0$.

The situation is quite different if the equations are interchanged prior to eliminating x_1:

$$3 x_1 + \quad x_2 = 1 \tag{4.3.17}$$
$$0 \cdot 001 x_1 - 2 x_2 = 4$$

Applying the Gaussian elimination procedure to these equations, and truncating the results to three decimal digits, gives

$$3 x_1 + \quad x_2 = 1 \tag{4.3.18}$$
$$-2 x_2 = 4$$

A correct value of $x_2 = -2$ is obtained from the second equation, and back-substituting this into the first equation gives the correct result, $x_1 = 1$.

This example demonstrates that too small a value for the pivot can lead to a serious loss of accuracy. It also suggests a way of avoiding these difficulties; the equation with the small pivot value should be interchanged with an equation having a larger pivot value. One further complication remains however. Multiplying any equation by a constant does not affect the solutions, so that equations (4.3.15) and the following set are equivalent:

$$10 x_1 - 20000 x_2 = 40000 \tag{4.3.19}$$
$$3 x_1 + \quad x_2 = 1$$

When x_1 is to be eliminated, the first equation then has the largest pivot value (that is, the largest coefficient of x_1). Nevertheless, as has been demonstrated, if this equation is used to perform the elimination, the result will be incorrect.

Normalization is one way of ensuring that the pivot is chosen correctly.

For each equation j, the value q_j of the largest coefficient is determined:

$$q_j = \max_{k=1}^{n} |A_{jk}| \qquad\qquad (4.3.20)$$

The equation can now be normalized by dividing it by q_j. However, normalizing the coefficients leads to additional rounding errors. Fortunately there is a strategy for selecting the best pivot without resorting to full normalization. When a variable x_i is to be eliminated the equations are searched for the highest pivot relative to the norm q_j. That is, the value of j is selected which maximizes $|A_{ji}|/q_j$. Equations i and j are then interchanged before x_i is eliminated.

Matrices are not, of course, predefined data types in Modula-2 and must therefore be declared before use:

```
TYPE
    index = [0..maxsize];
    vector = ARRAY index OF REAL;
    matrix = ARRAY index OF vector;
```

A procedure for solving a set of n simultaneous equations is shown below.

```
PROCEDURE solveq(a: matrix; VAR x: vector; b: vector; n: index);
VAR
    i, j, k: index;
    p: CARDINAL;
    t: REAL;
    q, s: vector;
BEGIN
    FOR i := 1 TO n DO
        t := 0.0;
        FOR j := 1 TO n DO
            IF ABS(a[i, j]) > t THEN t := ABS(a[i, j]) END
        END;
        q[i] := t
    END;
    FOR i := 1 TO n-1 DO
        t := 0.0;
        p := 0;
```

```
        FOR j := i TO n DO
            IF ABS(a[j, i]/q[j]) > t THEN
                t := ABS(a[j, i]/q[j]);
                p := j
            END
        END;
        IF p = 0 THEN
            WriteString('Matrix Singular');
            WriteLn;
            HALT
        END;
        IF p <> i THEN
            s := a[p]; a[p] := a[i]; a[i] := s;
            t := b[p]; b[p] := b[i]; b[i] := t
        END;
        FOR j := i+1 TO n DO
            t := a[j, i]/a[i, i];
            FOR k := i+1 TO n DO
                a[j, k] := a[j, k]-t*a[i, k]
            END;
            b[j] := b[j]-t*b[i]
        END
    END;
    FOR i := n TO 1 BY -1 DO
        t := b[i]/a[i, i];
        FOR j := 1 TO i-1 DO
            b[j] := b[j]-t*a[j, i]
        END;
        x[i] := t
    END
END solveq;
```

If, after searching for the best pivot, it is found to be zero, then the matrix of coefficients is singular and no solution to the equations exists. In this case a warning message is printed and the program halted.

An estimate of the efficiency of this procedure can be deduced from the fact that the number of nested FOR loops never exceeds three. The loop variables i, j and k each take no more than n values; consequently the loop,

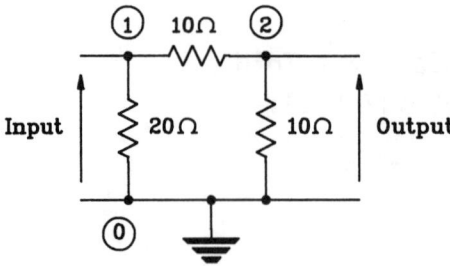

Figure 4.7
A simple
resistive
network

which contains one multiplication, is executed less than n^3 times. A more detailed analysis shows the number of multiplications to be $n^3/3$.

To see how numerical network analysis works in practice, consider the simple resistive network shown in figure 4.7. Applying the rules given in (4.2.20) to this network allows the Y-matrix to be evaluated:

$$\begin{bmatrix} 0 \cdot 15 & -0 \cdot 1 \\ -0 \cdot 1 & 0 \cdot 2 \end{bmatrix} \begin{bmatrix} V_1 \\ V_2 \end{bmatrix} = \begin{bmatrix} I_1 \\ I_2 \end{bmatrix} \tag{4.3.21}$$

Putting $I_1 = 1$ A (at the input node) and $I_2 = 0$ gives a set of simultaneous equations:

$$\begin{bmatrix} 0 \cdot 15 & -0 \cdot 1 \\ -0 \cdot 1 & 0 \cdot 2 \end{bmatrix} \begin{bmatrix} V_1 \\ V_2 \end{bmatrix} = \begin{bmatrix} 1 \\ 0 \end{bmatrix} \tag{4.3.22}$$

The solution to these equations is

$$\begin{aligned} V_1 &= 10\text{V} \\ V_2 &= 5\text{V} \end{aligned} \tag{4.3.23}$$

From these results it can be deduced that the input impedance is 10 Ω, the transimpedance is 5 Ω, and the voltage gain is 0·5.

4.4 Analysis of Reactive Networks

Most networks are not, of course, purely resistive but contain reactive elements, namely inductors and capacitors. This introduces two extra complications: the Y-matrix is complex and is frequency-dependent.

Frequency-dependence of the Y-matrix means that it must be set up, and the simultaneous equations solved, for each frequency of interest. A typical Bode plot might involve calculation of the gain and phase at one hundred separate frequencies. Clearly this requires a substantial computational effort.

Since the Y-matrix is complex, the nodal voltages must also be complex. For generality it will be assumed that the external currents are complex, although in practice they will usually be real. The complex simultaneous equations can be represented by

$$\mathbf{Y}_c \, \mathbf{V}_c = \mathbf{I}_c \tag{4.4.1}$$

where

$$\mathbf{Y}_c = \mathbf{Y}_r + j\mathbf{Y}_i$$
$$\mathbf{V}_c = \mathbf{V}_r + j\mathbf{V}_i \qquad (4.4.2)$$
$$\mathbf{I}_c = \mathbf{I}_r + j\mathbf{I}_i$$

One approach to solving these equations is to use the type *complex* defined in chapter 2, for elements of the matrix $\mathbf{Y}_c$ and the vectors $\mathbf{V}_c$, $\mathbf{I}_c$. The Gaussian elimination procedure can then be generalized to deal with complex variables. Each multiplication operation of the original procedure will require four multiplications in the complex procedure leading to a total of $4n^3/3$. There will be an additional computational overhead associated with accessing the real and imaginary parts of the variables.

An alternative method for solving the complex equations is to turn them into an equivalent set of real equations. To see how this can be done, write the variables of equation (4.4.1) in terms of their real and imaginary parts:

$$(\mathbf{Y}_r + j\mathbf{Y}_i)(\mathbf{V}_r + j\mathbf{V}_i) = \mathbf{I}_r + j\mathbf{I}_i \qquad (4.4.3)$$

Comparing real and imaginary parts of equation (4.4.3) yields a set of two matrix-vector equations:

$$\mathbf{Y}_r \mathbf{V}_r - \mathbf{Y}_i \mathbf{V}_i = \mathbf{I}_r$$
$$\mathbf{Y}_i \mathbf{V}_r + \mathbf{Y}_r \mathbf{V}_i = \mathbf{I}_i \qquad (4.4.4)$$

These equations can themselves be written in matrix-vector form:

$$\begin{bmatrix} \mathbf{Y}_r & -\mathbf{Y}_i \\ \mathbf{Y}_i & \mathbf{Y}_r \end{bmatrix} \begin{bmatrix} \mathbf{V}_r \\ \mathbf{V}_i \end{bmatrix} = \begin{bmatrix} \mathbf{I}_r \\ \mathbf{I}_i \end{bmatrix} \qquad (4.4.5)$$

or

$$\mathbf{Y}_2 \mathbf{V}_2 = \mathbf{I}_2 \qquad (4.4.6)$$

The new matrix $\mathbf{Y}_2$ is a $2n \times 2n$ real matrix; $\mathbf{V}_2$ and $\mathbf{I}_2$ are $2n \times 1$ real vectors. This set of real simultaneous equations can be solved by the normal Gaussian elimination procedure. There is an efficiency penalty however. Doubling the number of equations leads to an eightfold increase in the number of multiplications. Although this method requires $8n^3/3$ multiplications compared with $4n^3/3$ multiplications for the complex Gaussian elimination method, it may nevertheless be preferred on account of its simplicity.

Consider the simple high-pass filter illustrated in figure 4.8. At a frequency of 2×10^4 rad/s this network has a Y-matrix given by

$$\mathbf{Y}_c = \begin{bmatrix} j0 \cdot 02 & -j0 \cdot 02 \\ -j0 \cdot 02 & 0 \cdot 01 + j0 \cdot 02 \end{bmatrix} \qquad (4.4.7)$$

Figure 4.8
A first-order
high-pass filter

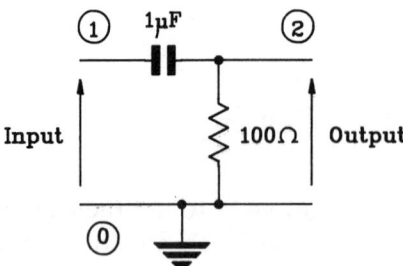

which can be converted into a 4×4 real matrix:

$$\mathbf{Y}_2 = \begin{bmatrix} 0 & 0 & -0\cdot02 & 0\cdot02 \\ 0 & 0\cdot01 & 0\cdot02 & -0\cdot02 \\ 0\cdot02 & -0\cdot02 & 0 & 0 \\ -0\cdot02 & 0\cdot02 & 0 & 0\cdot01 \end{bmatrix} \qquad (4.4.8)$$

In choosing an appropriate external current vector $\mathbf{I}_c$, it should be remembered that the value of the input generator I_t will cancel out of the final result, whether it be input impedance, transimpedance or voltage gain. Both the magnitude and the phase of the input generator are therefore quite arbitrary, and the simplest non-zero value of I_t should be used:

$$I_t = 1 + j0 \qquad (4.4.9)$$

This leads to a current vector $\mathbf{I}_c$ given by

$$\mathbf{I}_c = \begin{bmatrix} 1 + j0 \\ 0 \end{bmatrix} \qquad (4.4.10)$$

which can be converted to a real 4×1 current vector:

$$\mathbf{I}_2 = \begin{bmatrix} 1 \\ 0 \\ 0 \\ 0 \end{bmatrix} \qquad (4.4.11)$$

The matrix equation can now be solved by using a real Gaussian elimination method to give a 4×1 real vector of node voltages:

$$\mathbf{V}_2 = \begin{bmatrix} 100 \\ 100 \\ -50 \\ 0 \end{bmatrix} \qquad (4.4.12)$$

Finally, the 4×1 real vector is reduced to a 2×1 complex vector:

$$\mathbf{V}_c = \begin{bmatrix} 100 - j50 \\ 100 + j0 \end{bmatrix} \qquad (4.4.13)$$

or

$$\begin{aligned} V_1 &= 100 - j50 \\ V_2 &= 100 \end{aligned} \qquad (4.4.14)$$

From these results it can be deduced that at a frequency of 2×10^4 rad/s the input impedance is $100 - j50 \ \Omega$, the transimpedance is $100 \ \Omega$ and the voltage gain is $100/(100 - j50)$. Using the method described in chapter 2, the complex voltage gain can be converted to its magnitude and phase:

$$\text{Gain} = -0 \cdot 97 \ \text{dB}$$
$$\text{Phase} = 26 \cdot 5^\circ$$

(4.4.15)

4.5 Analysis of Active Networks

Nodal analysis can easily be extended to include voltage-controlled current sources (VCCS). Other types of controlled source are more difficult and will be discussed, together with infinite-gain operational amplifiers, in the following section. Fortunately, the most important linear active devices, namely bipolar transistors and field-effect transistors, are, in the small signal limit, good approximations to VCCS.

Figure 4.9 shows a four-terminal VCCS with input (or control) nodes i, j, and output nodes k, l. The current i_r flowing in the output is proportional to the voltage difference v_r across the input:

$$i_r = y_r v_r \tag{4.5.1}$$

where y_r is the transconductance of the VCCS. As far as the branch-admittance matrix $\mathbf{C}$ is concerned, therefore, the VCCS is similar to passive components: the diagonal element of $\mathbf{C}$ corresponding to the VCCS is simply equal to y_r.

The main additional complexity that results from the inclusion of VCCS is that a single network graph (and its associated incidence matrix) is no longer sufficient to represent the network topology. This is a consequence of the fact that the nodes (i, j) which control the branch voltage v_r of a VCCS are not the same as the nodes (k, l) into which the branch current i_r flows. The incidence matrix linking the branch voltages to the node voltages is therefore distinct from the incidence matrix expressing Kirchhoff's current law. To distinguish these matrices they will be termed the voltage incidence matrix, with symbol $\mathbf{A_v}$, and the current incidence matrix, with symbol $\mathbf{A_c}$. Equations (4.2.7) and (4.2.14) become

$$\mathbf{A_c \, i} = \mathbf{I} \tag{4.5.2}$$

$$\mathbf{A_v^T \, V} = \mathbf{v} \tag{4.5.3}$$

Figure 4.9
A four-terminal voltage-controlled current source

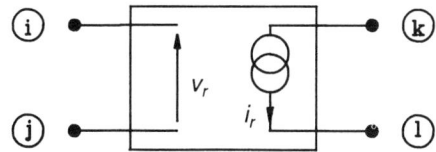

59

Figure 4.10
A parallel-
feedback
voltage
amplifier

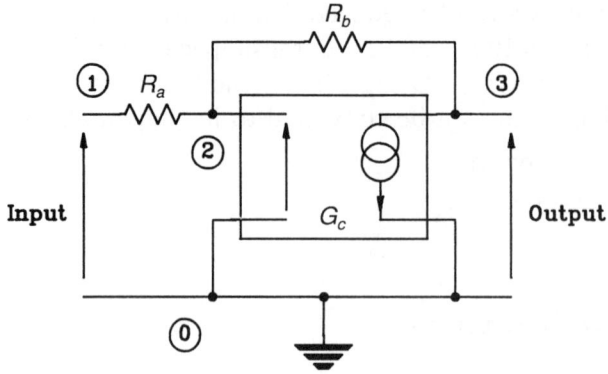

The Y-matrix is then given by

$$\mathbf{Y} = \mathbf{A_c}\, \mathbf{C}\, \mathbf{A_v}^T \qquad\qquad (4.5.4)$$

Each of the incidence matrices is associated with its own graph according to the rules given in (4.2.8). Two graphs, the voltage graph and the current graph, are therefore required to define the network topology of an active network.

As far as the passive components are concerned, the matrices $\mathbf{A_v}$ and $\mathbf{A_c}$ are identical. It follows that the voltage and current graphs for the passive components are also identical. In the case of VCCS, only the input nodes affect $\mathbf{A_v}$ and the voltage graph therefore has a branch corresponding to the input nodes. Also, only the output nodes affect $\mathbf{A_c}$ and the current graph therefore has a branch corresponding to the output nodes.

Consider the parallel feedback voltage amplifier shown in figure 4.10. The branch admittance matrix for this network is given by

$$\mathbf{C} = \begin{bmatrix} y_a & 0 & 0 \\ 0 & y_b & 0 \\ 0 & 0 & y_c \end{bmatrix} \qquad\qquad (4.5.5)$$

Figure 4.11 shows the two-graph representation of the topology of this

Figure 4.11
Graphs of the
active network

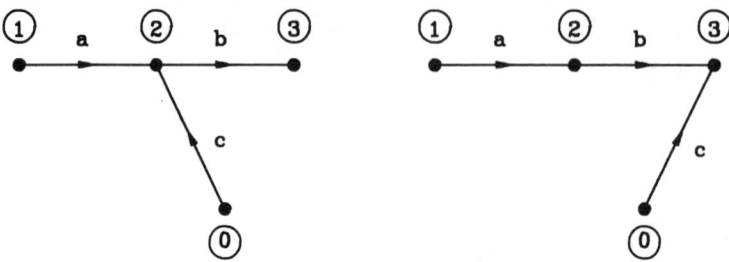

Voltage Graph   Current Graph

network. From these graphs the voltage and current incidence matrices can be derived:

$$\mathbf{A_v} = \begin{bmatrix} 1 & 0 & 0 \\ -1 & 1 & -1 \\ 0 & -1 & 0 \end{bmatrix} \tag{4.5.6}$$

$$\mathbf{A_c} = \begin{bmatrix} 1 & 0 & 0 \\ -1 & 1 & 0 \\ 0 & -1 & -1 \end{bmatrix} \tag{4.5.7}$$

Finally the branch admittance matrix and the incidence matrices can be combined according to equation (4.5.4) to give the Y-matrix:

$$\mathbf{Y} = \begin{bmatrix} y_a & -y_a & 0 \\ -y_a & y_a + y_b & -y_b \\ 0 & y_c - y_b & y_b \end{bmatrix} \tag{4.5.8}$$

If none of the input or output nodes of the VCCS is the reference node, then the VCCS will contribute to four elements of the Y-matrix. This will be reduced to two elements if one of the input nodes or one of the output nodes is ground, and will be reduced to a single element if one of the input nodes and one of the output nodes is ground.

The Y-matrix for the VCCS shown in figure 4.9 is given by

$$
\mathbf{Y} = \begin{matrix} \\ k \\ l \end{matrix} \overset{r}{\begin{bmatrix} 0 & 0 & 0 \\ 0 & 1 & 0 \\ 0 & -1 & 0 \\ 0 & 0 & 0 \end{bmatrix}} \begin{bmatrix} 0 & & \\ & y_r & \\ & & 0 \end{bmatrix} \overset{i \quad j}{\begin{bmatrix} 0 & 0 & 0 & 0 \\ 0 & 1 & -1 & 0 \\ 0 & 0 & 0 & 0 \end{bmatrix}} r
$$

$$
= \begin{matrix} k \\ l \end{matrix} \overset{i \qquad j}{\begin{bmatrix} y_r & -y_r \\ -y_r & y_r \end{bmatrix}} \tag{4.5.9}
$$

This VCCS contributes to the elements Y_{ki}, Y_{lj}, and with a negative sign to the elements Y_{kj}, Y_{li}.

It is possible to consider a two-terminal passive component as a special type of VCCS in which the input nodes are identical to the output nodes. In this way some simplification of the procedure for setting up the Y-matrix can be achieved.

Figure 4.12 shows the low-frequency small-signal equivalent circuit of a bipolar transistor voltage amplifier with series feedback. Using the rules that have been established for constructing the Y-matrix, and setting all the external current generators to zero with the exception of I_1, gives the nodal

61

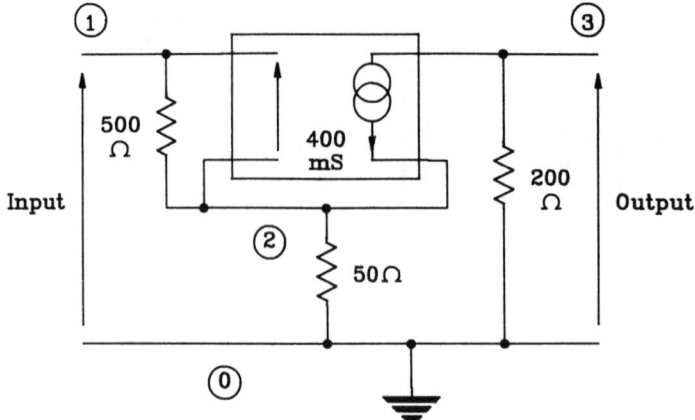

Figure 4.12
A transistor amplifier

equations for the network:

$$\begin{bmatrix} 0\cdot002 & -0\cdot002 & 0 \\ -0\cdot402 & 0\cdot422 & 0 \\ 0\cdot4 & -0\cdot4 & 0\cdot005 \end{bmatrix} \begin{bmatrix} V_1 \\ V_2 \\ V_3 \end{bmatrix} = \begin{bmatrix} 1 \\ 0 \\ 0 \end{bmatrix} \qquad (4.5.10)$$

These equations can be solved by Gaussian elimination to give the node voltages:

$$\begin{aligned} V_1 &= 10550 \\ V_2 &= 10050 \\ V_3 &= 40000 \end{aligned} \qquad (4.5.11)$$

From these results it can be deduced that the input impedance is $10\cdot55$ kΩ and the voltage gain is $3\cdot79$.

It is possible to use a VCCS, together with some passive components, to approximate other types of controlled source. For example, a VCCS with a high value conductance connected across its output terminals can be used to approximate a voltage-controlled voltage source. This method must be employed with some care, however. A good approximation is only obtained if the auxiliary conductance is much greater than any other conductances connected to the output of the VCCS. This can lead to the set of nodal equations being poorly conditioned (that is, near to being singular) with consequent adverse affects on the resultant accuracy.

4.6 Networks Containing Operational Amplifiers

It will be assumed that the operational amplifiers are ideal and can be treated as infinite-gain voltage-controlled voltage sources. For generality, four-terminal operational amplifiers, as shown in figure 4.13 will be considered. Actual operational amplifiers have only one accessible output terminal; the other output terminal has an implicit connection to ground via the power supplies.

62

Figure 4.13
A four-terminal
operational
amplifier

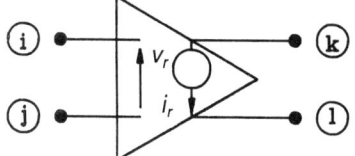

Figure 4.14
A series-
feedback
voltage
amplifier

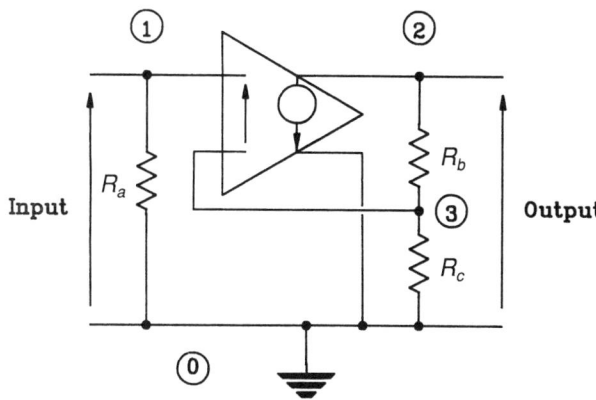

When an infinite-gain operational amplifier is embedded in a passive network, it will have two effects. Since the output of the amplifier is a voltage generator, the output branch current i_r is indeterminate. Also, the infinite voltage gain constrains the input branch voltage v_r to be zero.

To see how the network graphs and matrices are affected by the presence of an operational amplifier, consider the series-feedback amplifier shown in figure 4.14.

If the operational amplifier is excluded, then the topology of the remaining passive network can be represented by the voltage and current graphs (which are identical for a passive network) shown in figure 4.15.

The passive branch currents and branch voltages obey Ohm's law:

$$
\begin{bmatrix} i_a \\ i_b \\ i_c \end{bmatrix} = \begin{bmatrix} y_a & 0 & 0 \\ 0 & y_b & 0 \\ 0 & 0 & y_c \end{bmatrix} \begin{bmatrix} v_a \\ v_b \\ v_c \end{bmatrix}
\tag{4.6.1}
$$

Figure 4.15
Graphs of the
passive
network

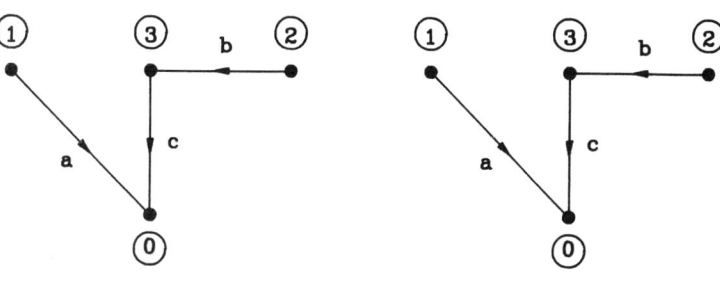

Voltage Graph   Current Graph

63

Kirchhoff's current law can be applied to each of the nodes to give the set of equations shown below:

$$
\begin{bmatrix} 1 & 0 & 0 \\ 0 & 1 & 0 \\ 0 & -1 & 1 \end{bmatrix} \begin{bmatrix} i_a \\ i_b \\ i_c \end{bmatrix} = \begin{bmatrix} I_1 \\ I_2 \\ I_3 \end{bmatrix}
\tag{4.6.2}
$$

One effect of the operational amplifier is to invalidate the equation for I_2 because an indeterminate current flows into node 2 from the output of the amplifier. This equation is therefore eliminated, leaving

$$
\begin{bmatrix} 1 & 0 & 0 \\ 0 & -1 & 1 \end{bmatrix} \begin{bmatrix} i_a \\ i_b \\ i_c \end{bmatrix} = \begin{bmatrix} I_1 \\ I_3 \end{bmatrix}
\tag{4.6.3}
$$

In general, if the output nodes of the operational amplifier are k and l, then the individual equations for I_k and I_l are no longer valid. However, if the equations for I_k and I_l are added together, then the contributions from the output current i_r of the amplifier cancel out. The result is a single equation covering the two nodes k and l. This is equivalent to coalescing the nodes k and l on the current graph for the network; any existing branches between k and l are eliminated.

The effect on the current incidence matrix $\mathbf{A_c}$ depends on whether either of k and l is equal to the reference node. If k is the reference node, then the row of $\mathbf{A_c}$ corresponding to node l is deleted; if l is the reference node, then the row corresponding to node k is deleted; otherwise rows k and l are combined by addition to give a single row.

The node voltages V_i of the passive network are related to the branch voltages v_j by the set of equations shown below:

$$
\begin{bmatrix} 1 & 0 & 0 \\ 0 & 1 & -1 \\ 0 & 0 & 1 \end{bmatrix} \begin{bmatrix} V_1 \\ V_2 \\ V_3 \end{bmatrix} = \begin{bmatrix} v_a \\ v_b \\ v_c \end{bmatrix}
\tag{4.6.4}
$$

As a consequence of the infinite gain of the amplifier, the voltage v_r across its input terminals must be zero. In other words $V_3 = V_1$. It is therefore possible to add together columns 1 and 3 of the matrix to give

$$
\begin{bmatrix} 1 & 0 \\ -1 & 1 \\ 1 & 0 \end{bmatrix} \begin{bmatrix} V_1 \\ V_2 \end{bmatrix} = \begin{bmatrix} v_a \\ v_b \\ v_c \end{bmatrix}
\tag{4.6.5}
$$

In general, if the input nodes of the operational amplifier are i and j, then $V_i = V_j$, and V_i may therefore be replaced by V_j in the equations linking the node and branch voltages. This is equivalent to coalescing nodes i and j on the voltage graph for the network; any existing branches between i and j are eliminated.

Figure 4.16
Graphs of the
network
including the
amplifier

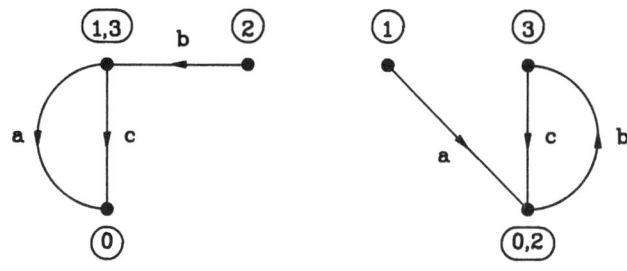

Voltage Graph   **Current Graph**

Figure 4.16 shows the graphs of the network modified to take account of the effects of the operational amplifier. The effect on the voltage incidence matrix $\mathbf{A_v}$ depends on the whether either i or j is equal to the reference node. If i is the reference node, then the row of $\mathbf{A_v}$ corresponding to node j is deleted; if j is the reference node, then the row corresponding to node i is deleted; otherwise rows i and j are combined by addition to give a single row.

Finally, equations (4.6.1), (4.6.3) and (4.6.5) can be combined to give the nodal equations for the complete network:

$$\begin{bmatrix} 1 & 0 & 0 \\ 0 & -1 & 1 \end{bmatrix} \begin{bmatrix} y_a & 0 & 0 \\ 0 & y_b & 0 \\ 0 & 0 & y_c \end{bmatrix} \begin{bmatrix} 1 & 0 \\ -1 & 1 \\ 1 & 0 \end{bmatrix} \begin{bmatrix} V_1 \\ V_2 \end{bmatrix} = \begin{bmatrix} I_1 \\ I_3 \end{bmatrix} \tag{4.6.6}$$

and the Y-matrix is therefore given by

$$\mathbf{Y} = \begin{bmatrix} y_a & 0 \\ y_b + y_c & -y_b \end{bmatrix} \tag{4.6.7}$$

Deleting or adding rows of the current incidence matrix will have a similar effect on the resultant Y-matrix. Also, deleting or adding columns of the transposed voltage matrix will have a similar effect on the resultant Y-matrix. In general, when an operational amplifier is embedded in a network, the Y-matrix must be modified in the following way. If either of the output nodes is the reference node, then the row corresponding to the other output node is deleted, otherwise the rows corresponding to the output nodes are combined by addition. If either of the input nodes is the reference node, then the column corresponding to the other input node is deleted, otherwise the columns corresponding to the input nodes are combined by addition. Each operational amplifier in the network reduces the size of the Y-matrix by one.

To see how this works in practice, consider the parallel-feedback voltage amplifier shown in figure 4.17. The nodal equations for the passive network,

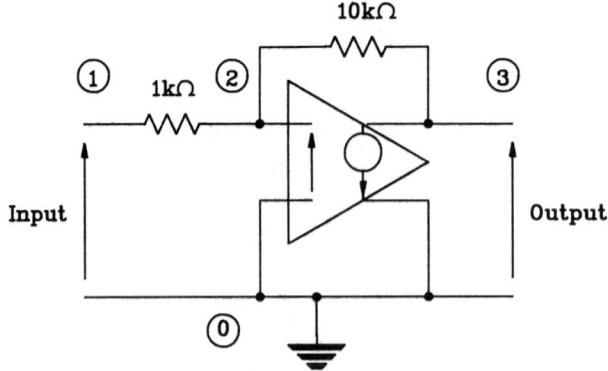

Figure 4.17
A parallel-
feedback
voltage
amplifier

excluding the operational amplifier are given by

$$
\begin{bmatrix}
0\cdot001 & -0\cdot001 & 0 \\
-0\cdot001 & 0\cdot0011 & -0\cdot0001 \\
0 & -0\cdot0001 & 0\cdot0001
\end{bmatrix}
\begin{bmatrix}
V_1 \\
V_2 \\
V_3
\end{bmatrix}
=
\begin{bmatrix}
I_1 \\
I_2 \\
I_3
\end{bmatrix}
\qquad (4.6.8)
$$

Now the Y-matrix is modified to take account of the presence of the operational amplifier. The row corresponding to the output node 3 of the amplifier is deleted, and the column corresponding to the input node 2 of the amplifier is deleted:

$$
\begin{bmatrix}
0\cdot001 & 0 \\
-0\cdot001 & -0\cdot0001
\end{bmatrix}
\begin{bmatrix}
V_1 \\
V_3
\end{bmatrix}
=
\begin{bmatrix}
I_1 \\
I_2
\end{bmatrix}
\qquad (4.6.9)
$$

Setting $I_1 = 1A$ and $I_2 = 0$ allows the equations to be solved and the nodal voltages determined:

$$
\begin{aligned}
V_1 &= 1000 \\
V_3 &= -10\,000
\end{aligned}
\qquad (4.6.10)
$$

From these results the input impedance is found to be 1 kΩ and the voltage gain to be -10.

4.7 Limitations of Numerical Network Analysis

Perhaps the most serious limitation of numerical network analysis is the amount of computation required. If the network is reactive, as is nearly always the case, then the response will need to be evaluated at a number of different frequencies. Each evaluation involves solving a set of simultaneous linear equations. To give some idea of how serious a problem this might be, the network shown in figure 4.18 was analyzed numerically. The network, which consists of 14 nodes and 15 passive components represents a fifth-order high-pass filter.

In order to provide an adequate display of the frequency response of this filter, the gain must be evaluated at around one hundred separate

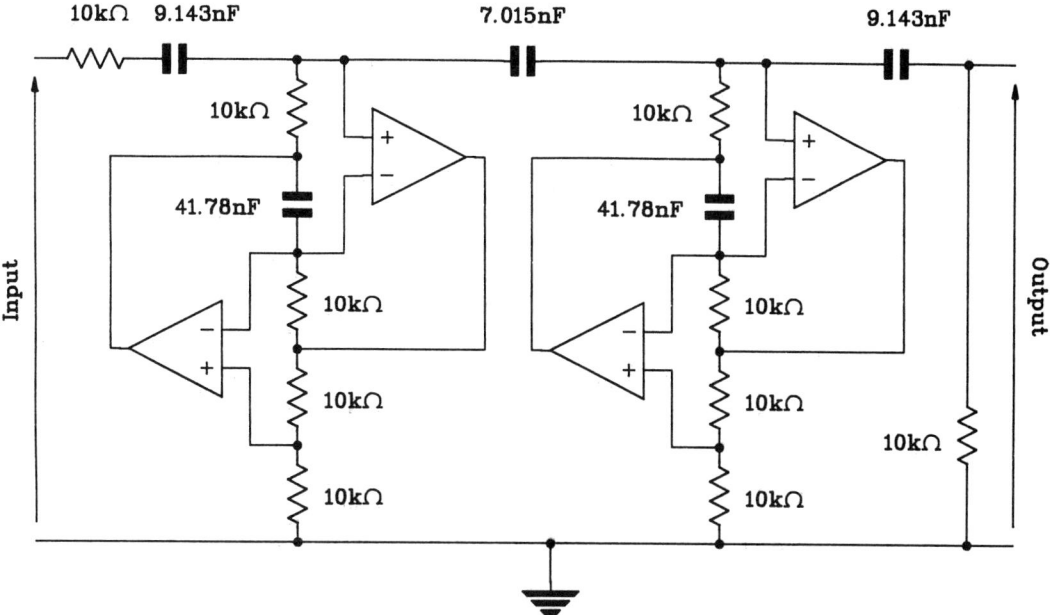

Figure 4.18 A fifth-order Chebychev high-pass filter

frequencies. Using a standard IBM personal computer, each evaluation took about 4 sec. Of course, most personal computers are more powerful than the standard IBM-PC, typically by a factor of between 2 and 20. Nevertheless, this represents a substantial amount of computational effort.

A problem that can sometimes arise in network analysis by the nodal admittance method is that a poorly-conditioned Y-matrix can lead to a loss of accuracy. Unfortunately this can happen, even for perfectly reasonable networks, if the admittances of the components differ widely. The loss of accuracy results from the subtraction of nearly equal terms during the Gaussian elimination process. To see how a poorly-conditioned Y-matrix can arise, consider the potential divider shown in figure 4.19. The nodal equations for this network are

$$\begin{bmatrix} 1 & -1 \\ -1 & 1 \cdot 000001 \end{bmatrix} \begin{bmatrix} V_1 \\ V_2 \end{bmatrix} = \begin{bmatrix} I_1 \\ I_2 \end{bmatrix} \tag{4.7.1}$$

Figure 4.19
A potential
divider

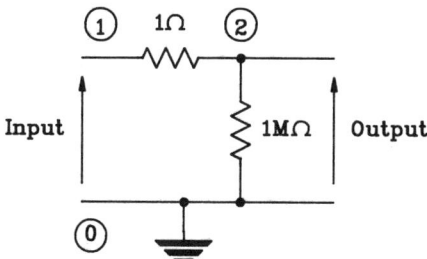

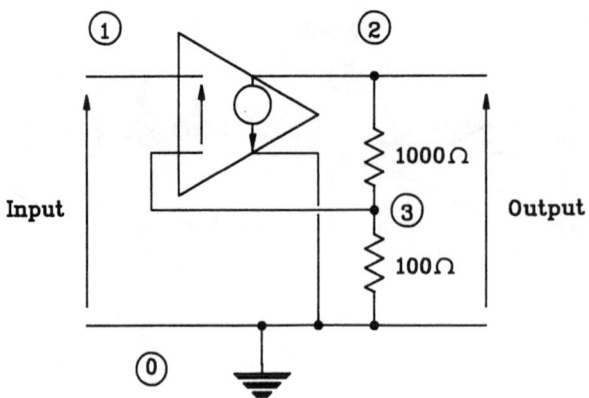

Figure 4.20
A series-feedback voltage amplifier

Input

$1000\,\Omega$

Output

$100\,\Omega$

By adding the first equation to the second, the variable V_1 can be eliminated:

$$\begin{bmatrix} 1 & -1 \\ 0 & 0\cdot000001 \end{bmatrix}\begin{bmatrix} V_1 \\ V_2 \end{bmatrix} = \begin{bmatrix} I_1 \\ I_1 + I_2 \end{bmatrix} \tag{4.7.2}$$

Although this new set of equations is apparently correct, the lower right-hand element of the matrix was obtained by subtraction of relatively much larger quantities, and six digits of precision have been lost. Since the voltage gain depends directly on this element of the matrix, it will suffer a similar loss of precision.

Finally, in the case of active networks, care should be taken to ensure that the input conductance is non-zero, otherwise the matrix will be singular and the analysis will fail. Figure 4.20 shows a series-feedback voltage amplifier with zero input conductance. The nodal equations for this network are

$$\begin{bmatrix} 0 & 0 \\ 0\cdot011 & -0\cdot001 \end{bmatrix}\begin{bmatrix} V_1 \\ V_2 \end{bmatrix} = \begin{bmatrix} I_1 \\ I_3 \end{bmatrix} \tag{4.7.3}$$

Obviously there is no solution to these equations. Fortunately, the problem can be overcome by a very simple expedient. An additional conductance, of similar magnitude to other conductances within the network, is connected from the input node to ground. This removes the singularity of the matrix while leaving the voltage gain of the network unaffected.

4.8 Summary

Numerical network analysis is used to obtain network functions, such as the voltage transfer ratio, as complex numbers. Although symbolic network functions contain more information than numerical network functions, symbolic analysis is more complex, and in some circumstances less efficient than numerical analysis.

In nodal analysis it is assumed that external currents flow into the network nodes. Kirchhoff's current law is applied to each of the nodes, except the reference node, to give a set of simultaneous linear equations in the nodal voltages, from which network functions such as the input impedance, the transimpedance and the voltage transfer ratio can be derived.

Formally the nodal equations are set up with the aid of network graphs which define completely the topology of a network. An alternative representation of network topology is the incidence matrix, which is entirely equivalent to, and can be derived from, the network graph. Combining the incidence matrix with the branch admittance matrix, which is a diagonal matrix of the individual component admittances, gives the nodal admittance matrix or Y-matrix. The Y-matrix, together with the vector of external currents constitutes the nodal equations.

Simultaneous linear equations are usually solved by Gaussian elimination. Care must be taken during Gaussian elimination to choose the correct pivot; failure to do so can lead to a serious loss of accuracy.

In the case of reactive networks the Y-matrix is complex and frequency-dependent. The nodal equations must therefore be set up, and solved, for each frequency of interest. Since the equations are complex they must either be converted to a larger set of real equations, or be solved by a complex Gaussian elimination procedure.

Nodal analysis can be extended to deal with active networks containing voltage-controlled current sources or infinite-gain operational amplifiers. Two graphs, the voltage graph and the current graph, are necessary to define the topology of active networks. The voltage and current incidence matrices, which correspond to the voltage and current graphs, can be combined to give the Y-matrix of an active network.

Under some circumstances the Y-matrix may be poorly conditioned. This can happen with perfectly reasonable networks if there is a wide spread of admittance values, and may result in a loss of accuracy.

4.9 Problems

1 Draw the network graph of the passive attenuator shown in figure 4.21. Use this network graph to derive the incidence matrix $\mathbf{A}$, and hence the nodal admittance matrix $\mathbf{Y}$. Set all elements of the external current vector $\mathbf{I}$ to zero except $I_1 = 1\text{A}$ and solve the nodal equations:

$$\mathbf{Y}\,\mathbf{V} = \mathbf{I}$$

to obtain the vector of nodal voltages $\mathbf{V}$. Deduce the voltage gain V_2/V_1 and the input impedance V_1/I_1 of the network.

2 The network shown in figure 4.22 represents a twin-T notch filter. Draw the network graph of the twin-T filter and use this to derive the incidence

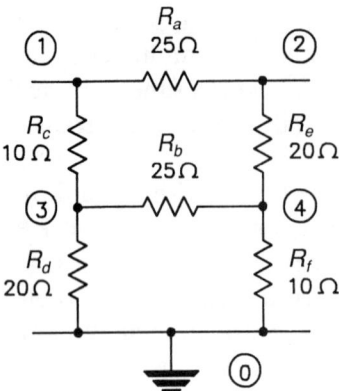

Figure 4.21
A passive attenuator network

R_a 25 Ω

R_c 10 Ω R_b 25 Ω R_e 20 Ω

R_d 20 Ω R_f 10 Ω

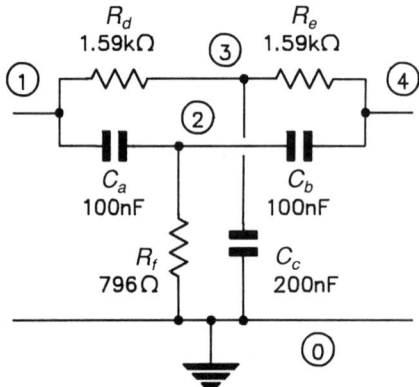

Figure 4.22
A twin-T filter

R_d 1.59 kΩ R_e 1.59 kΩ

C_a 100 nF C_b 100 nF

R_f 796 Ω C_c 200 nF

matrix **A**. Taking the frequency to be 500 Hz, calculate the 4 × 4 complex nodal admittance matrix $\mathbf{Y}_c$. Set all elements of the complex external current vector $\mathbf{I}_c$ to zero except $I_{c1} = 1 + j0$ A and solve the nodal equations:

$$\mathbf{Y}_c\mathbf{V}_c = \mathbf{I}_c$$

to obtain the vector of complex nodal voltages $\mathbf{V}_c$. Deduce the voltage gain $|V_4/V_1|$ and the phase shift of the network. Repeat the calculation for a range of frequencies between 100 Hz and 10 kHz.

3 Figure 4.23 shows the small-signal equivalent circuit of a two-transistor voltage amplifier with overall negative feedback. Draw the voltage and current graphs of this network, and use them to derive the voltage incidence matrix $\mathbf{A}_v$, and the current incidence matrix $\mathbf{A}_c$. Hence obtain the nodal admittance matrix $\mathbf{Y}$. Solve the nodal equations to determine the voltage gain V_3/V_1 of the network.

4 The network shown in figure 4.24 represents a voltage-to-current converter driving a load resistance R_l. The network incorporates an infinite-gain operational amplifier. Draw the voltage and current graphs of this network, and use them to derive the voltage incidence matrix $\mathbf{A}_v$, and the

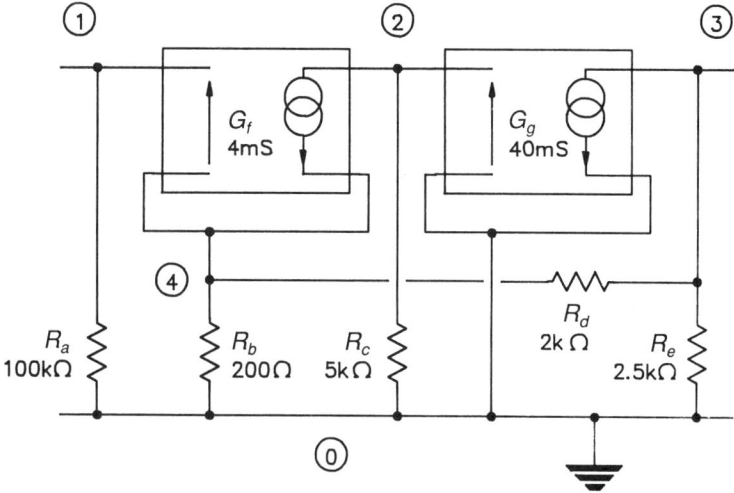

Figure 4.23
Equivalent circuit of a two-transistor voltage amplifier

G_f 4mS

G_g 40mS

R_a 100kΩ

R_b 200Ω

R_c 5kΩ

R_d 2kΩ

R_e 2.5kΩ

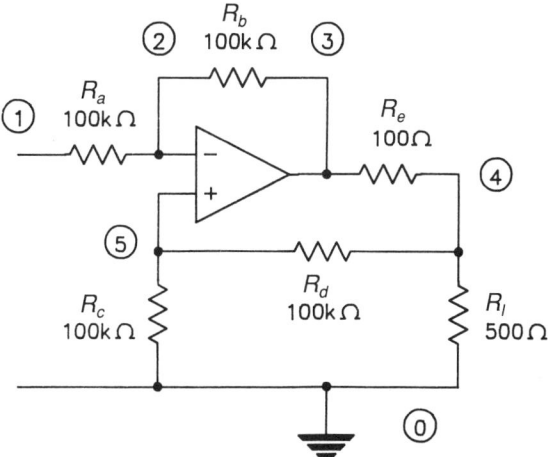

Figure 4.24
A voltage-to-current converter

R_b 100kΩ

R_a 100kΩ

R_e 100Ω

R_c 100kΩ

R_d 100kΩ

R_l 500Ω

current incidence matrix $\mathbf{A_c}$. Hence obtain the nodal admittance matrix $\mathbf{Y}$. Solve the nodal equations to determine the voltage gain V_4/V_1 of the network. By varying R_l show that the voltage gain is directly proportional to the load resistance.

5 The matrices given below are the voltage incidence matrix $\mathbf{A_v}$ and the current incidence matrix $\mathbf{A_c}$ of a network containing five passive admittances and two voltage-controlled current sources.

$$
\mathbf{A_v} = \begin{array}{c} \\ 1 \\ 2 \\ 3 \\ 4 \end{array}
\begin{array}{cccccccc}
a & b & c & d & e & f & g \\
\left[\begin{array}{ccccccc}
1 & 0 & 0 & 0 & 0 & 0 & 0 \\
-1 & 1 & 1 & 0 & 0 & 1 & 0 \\
0 & 0 & 0 & 1 & 0 & 0 & 1 \\
0 & 0 & -1 & 0 & 1 & 0 & -1
\end{array}\right]
\end{array}
$$

$$\mathbf{A_c} = \begin{array}{c} \\ 1 \\ 2 \\ 3 \\ 4 \end{array} \begin{array}{ccccccc} a & b & c & d & e & f & g \\ \left[\begin{array}{ccccccc} 1 & 0 & 0 & 0 & 0 & 0 & 0 \\ -1 & 1 & 1 & 0 & 0 & 0 & 0 \\ 0 & 0 & 0 & 1 & 0 & 1 & 0 \\ 0 & 0 & -1 & 0 & 1 & 0 & -1 \end{array}\right] \end{array}$$

Draw the current and voltage graphs of the network, and hence deduce the network diagram.

5 Symbolic Network Analysis

5.1 Introduction

Symbolic network analysis is concerned with determining the network functions of electrical networks in the form of rational functions of $j\omega$, where the coefficients are expressions which contain symbols representing the network elements. Consider the filter network shown in figure 5.1.

This network has a symbolic voltage transfer function given by:

$$H(j\omega) = \frac{R_e}{R_a + R_e + (L_b + R_aC_cR_e + R_aC_dR_e)j\omega + (L_bC_dR_e + R_aC_cL_b)(j\omega)^2 + R_aC_cL_bC_dR_e(j\omega)^3}$$

(5.1.1)

As shown in the introduction to the previous chapter, symbolic network functions can be reduced to complex numbers by substitution of the component values and the frequency, allowing, for example, the gain and phase to be determined. This is, however, a one-way process. Numerical network analysis generates a table of values against frequency for the particular network function, but it is impracticable to convert these results to symbolic form.

There are many reasons for wanting network functions in symbolic form. One important advantage of symbolic analysis is that it does not suffer from the accuracy problems that are inherent in numerical analysis. In the previous chapter it was demonstrated that even for perfectly reasonable networks, the Y-matrix may be poorly conditioned if the admittances in the

Figure 5.1
A passive low-pass ladder filter

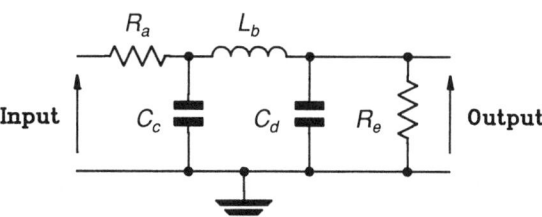

73

network differ widely. This leads to an inevitable loss of accuracy, as a result of the subtraction of nearly equal terms, during the process of solving the nodal equations. Suppose that the following expression is required to be evaluated as part of the solution of the nodal equations:

$$(r_1 r_2 + r_3 r_4) - r_1 r_2$$

If the term $r_1 r_2$ is larger in magnitude by a factor of 1000 than the term $r_3 r_4$, then the accuracy of the result will be three digits less than the full floating-point accuracy. This loss of accuracy never occurs when network functions are evaluated symbolically. Most symbolic analysis methods do not generate cancelling terms; even if such terms were to be generated, however, they could be cancelled symbolically with no loss of accuracy.

Symbolic network functions can be used to determine the sensitivities of various aspects of system performance to changes in component values. This is important in tolerance analysis where it may be required, for example, to determine by how much the gain at a particular frequency varies in response to changes in component values. Numerical methods for determining sensitivities are available. However, symbolic sensitivity functions have the advantage that the various factors contributing to a particular sensitivity function can be isolated.

Computer optimization of electronic networks has become an important design tool in recent years. Optimization is an iterative process and may involve many hundreds or thousands of evaluations of some network property (such as the voltage gain) at different frequencies, and with different component values. Clearly, for any but trivial networks, the computing effort involved in solving the nodal equations for each evaluation would be enormous. On the other hand, if the network function is available in symbolic form, then each evaluation simply involves substitution of the frequency and component values.

The time-domain response of a linear network can be obtained from the transfer function by, for example, performing an inverse Laplace transform. Although the time-domain response can also be obtained by performing a Fourier transform on a table of gain and phase, this approach involves a far greater computational effort and is less accurate.

Another important application of symbolic network functions is in determining the stability of a system. The poles of a network function determine its stability; if any pole of a system has a positive real part then the system is unstable. No satisfactory method is available for determining the stability from the results of numerical network analysis.

Finally, it often happens that a particular transfer function is required to be realized, and a suitable network configuration exists, but the component values are not known. If the transfer function of the network is available in symbolic form then the powers of jω in the numerators and denominators of the network transfer function and the required transfer function can be equated. The component values can then be obtained by solving the resultant simultaneous non-linear equations.

In spite of the many advantages of symbolic analysis it does have one serious drawback. The amount of computational effort required to generate a symbolic network function tends to increase exponentially with the number of nodes. An exact formula cannot be given because the complexity depends also on the number of branches and on the network configuration. By comparison, the computational effort required for numerical analysis increases as the cube of the number of nodes. For large networks, therefore, numerical analysis will be much more efficient and should be used wherever possible.

Before proceeding further it is worth considering why the Gaussian elimination procedure is unsuitable for symbolic analysis. A symbolic Gaussian elimination procedure would require each of the n^2 elements of the Y-matrix to be polynomials in the component symbols and jω. Storage is therefore likely to be a problem. More seriously, whenever two symbolic elements are multiplied or subtracted the result may contain cancelling terms. A time-consuming search must be performed after each of the $n^3/3$ operations to recognize such terms and remove them. The amount of computation involved rules out this approach, except for very simple networks.

In the previous chapter it was shown that the voltage V_s on a node s resulting from an external current I_t applied to a node t, is given by

$$V_s = [\mathbf{Y}^{-1}]_{st} I_t \tag{5.1.2}$$

$$= \frac{\Delta_{ts}}{\Delta} I_t$$

The method of analysis that will be described here involves symbolic evaluation of the determinant Δ and the cofactor Δ_{ts} of the Y-matrix. It makes use of the important concept of the trees of the network graphs.

5.2 Trees of the Network Graphs

A tree (or more strictly a spanning tree) of a graph is a subgraph which connects together all of the nodes of the original graph while generating no loops. Branch orientations are ignored for the purpose of determining the trees. Consider, for example, the graph shown in figure 5.2. The five spanning trees of this graph are illustrated in figure 5.3.

Trees are usually specified by the products of their branches; the trees of

Figure 5.2
A network
graph

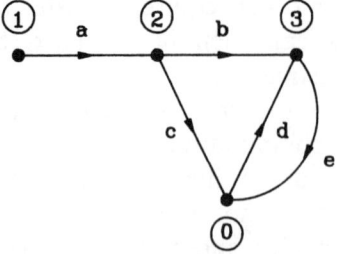

Figure 5.3
Spanning trees
of the network
graph

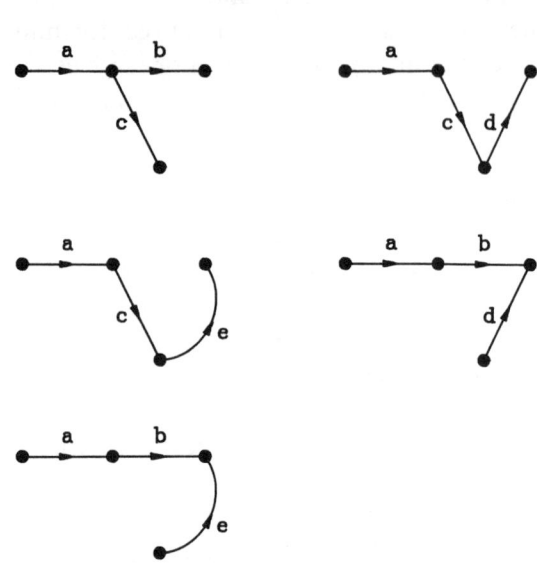

the graph shown in figure 5.2 can therefore be represented by:

abc
abd
abe
acd
ace

The number of branches in a spanning tree is determined by the number of nodes in the graph. A graph with two nodes will have trees which consist of a single branch. If a third node is added, then a second branch will be required to form a tree. Clearly this argument can be continued: each additional node needs an extra branch to connect it into the tree. A graph with r nodes will therefore have trees consisting of $(r-1)$ branches.

Spanning trees of a network graph are important in network analysis because they correspond directly with the non-singular majors of the incidence matrix. The significance of this fact will become apparent later. (A major of a rectangular matrix is the largest square submatrix. Thus the majors of an $r \times c$ matrix, where $r < c$, are the $r \times r$ matrices obtained by selecting r columns of the original matrix. If $r > c$ then the majors are the $c \times c$ matrices obtained by selecting c rows of the original matrix.)

The size of any incidence matrix $\mathbf{A}$ is $n \times m$, where n is the number of nodes excluding the reference node, and m is the number of branches. Consider the network graph shown in figure 5.2. This graph has an incidence matrix $\mathbf{A}$ given by

$$
\mathbf{A} = \begin{array}{c} \\ 1 \\ 2 \\ 3 \end{array}
\begin{array}{cccccc}
a & b & c & d & e \\
\left[\begin{array}{ccccc}
1 & 0 & 0 & 0 & 0 \\
-1 & 1 & 1 & 0 & 0 \\
0 & -1 & 0 & -1 & 1
\end{array}\right]
\end{array}
\tag{5.2.1}
$$

The transpose of $\mathbf{A}$ connects the node voltages V_i with the branch voltages v_j:

$$
\begin{array}{c} a \\ b \\ c \\ d \\ e \end{array}
\begin{array}{ccc} 1 & 2 & 3 \end{array}
\left[\begin{array}{ccc}
1 & -1 & 0 \\
0 & 1 & -1 \\
0 & 1 & 0 \\
0 & 0 & -1 \\
0 & 0 & 1
\end{array}\right]
\left[\begin{array}{c} V_1 \\ V_2 \\ V_3 \end{array}\right]
=
\left[\begin{array}{c} v_a \\ v_b \\ v_c \\ v_d \\ v_e \end{array}\right]
\tag{5.2.2}
$$

or

$$\mathbf{B}\,\mathbf{V} = \mathbf{v} \tag{5.2.3}$$

where $\mathbf{B} = \mathbf{A}^T$.

Suppose that a tree of the network graph, for example $\{abd\}$, is selected. Since the number of branches in a tree is one less than the number of nodes (including the reference node), there must be n branches in the tree. The matrix $\mathbf{B}$ can be partitioned into an $n \times n$ major whose rows correspond to the branches of the tree, and a matrix of the remaining rows:

$$
\begin{array}{c} a \\ b \\ d \\ \hline c \\ e \end{array}
\begin{array}{ccc} 1 & 2 & 3 \end{array}
\left[\begin{array}{ccc}
1 & -1 & 0 \\
0 & 1 & -1 \\
0 & 0 & -1 \\
\hline
0 & 1 & 0 \\
0 & 0 & 1
\end{array}\right]
\left[\begin{array}{c} V_1 \\ V_2 \\ V_3 \end{array}\right]
=
\left[\begin{array}{c} v_a \\ v_b \\ v_d \\ \hline v_c \\ v_e \end{array}\right]
\tag{5.2.4}
$$

or

$$
\left[\begin{array}{c} \mathbf{B}_t \\ \mathbf{B}_l \end{array}\right] \mathbf{V} = \left[\begin{array}{c} \mathbf{v}_t \\ \mathbf{v}_l \end{array}\right]
\tag{5.2.5}
$$

Now consider the tree partition alone:

$$\mathbf{B}_t \mathbf{V} = \mathbf{v}_t \tag{5.2.6}$$

The node voltages $\mathbf{V}$ are independent; so also are the branch voltages $\mathbf{v}_t$ in the tree because there are by definition no loops in the tree. It is therefore possible to invert the tree partition $\mathbf{B}_t$:

$$\mathbf{V} = \mathbf{B}_t^{-1} \mathbf{v}_t \tag{5.2.7}$$

From the definition of the incidence matrix all of the elements of $\mathbf{B}_t$ are either ± 1 or 0. Since any node voltage V_i can be expressed as a sum or difference of branch voltages v_j, the elements of $\mathbf{B}_t^{-1}$ must also be ± 1 or 0.

Now, the elements of $\mathbf{B}_t^{-1}$ are derived from the ratios of the cofactors of $\mathbf{B}_t$ to the determinant of $\mathbf{B}_t$. All of the non-zero elements of $\mathbf{B}_t^{-1}$ are ± 1 so that the cofactors and determinant of $\mathbf{B}_t$ must equal the same integer. By a similar argument the cofactors and determinant of $\mathbf{B}_t^{-1}$ must equal the same integer (although not necessarily the same integer as the determinant of $\mathbf{B}_t$). The product of the determinant of any matrix, and the determinant of its inverse is equal to unity:

$$|\mathbf{B}_t||\mathbf{B}_t^{-1}| = 1 \tag{5.2.8}$$

It follows from this equation, and from the fact that the determinants must be integer, that

$$|\mathbf{B}_t| = \pm 1 \tag{5.2.9}$$

or

$$|\mathbf{A}_t^T| = \pm 1 \tag{5.2.10}$$

In other words, the determinant of a major of the admittance matrix which corresponds with a tree of the network graph has a value ± 1.

Suppose now that the matrix $\mathbf{B}$ is partitioned into an $n \times n$ major which does not correspond to a tree of the network graph. The corresponding subgraph must therefore contain at least one loop, and at least one node must be disconnected. For example, suppose that the partition contains the rows b, c and d:

$$
\begin{array}{c}
 \\
b \\
c \\
d \\
\\
a \\
e
\end{array}
\begin{array}{ccc}
1 & 2 & 3
\end{array}
\left[
\begin{array}{ccc}
0 & 1 & -1 \\
0 & 1 & 0 \\
0 & 0 & 1 \\
\hline
1 & -1 & 0 \\
0 & 0 & 1
\end{array}
\right]
\left[
\begin{array}{c}
V_1 \\
V_2 \\
V_3
\end{array}
\right]
=
\left[
\begin{array}{c}
v_b \\
v_c \\
v_d \\
\\
v_a \\
v_e
\end{array}
\right]
\tag{5.2.11}
$$

The corresponding subgraph is shown in figure 5.4. Since one of the nodes is no longer connected, its voltage cannot appear in the expressions for the branch voltages; the column in the major partition of $\mathbf{B}$ corresponding to the disconnected node must therefore consist of zeros. It follows that the determinant of this major must be zero. Since $\mathbf{B}$ is the transpose of the

Figure 5.4
A non-tree
subgraph

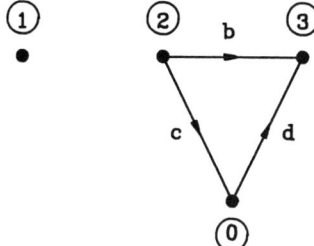

incidence matrix **A**, the major partition of **A** must also have a zero determinant.

The following important result follows from these considerations: a major of an incidence matrix **A** has a determinant equal to ± 1 provided that the major corresponds to a spanning tree of the network graph, otherwise the determinant is zero.

5.3 Determinant of the Y-Matrix

In the previous chapter the following result was derived:

$$\mathbf{Y} = \mathbf{A_c} \, \mathbf{C} \, \mathbf{A_v^T} \tag{5.3.1}$$

where $\mathbf{A_v}$ and $\mathbf{A_c}$ are the voltage and current incidence matrices, and $\mathbf{C}$ is the branch admittance matrix. If n is the number of nodes excluding the reference node, and m is the number of branches, then $\mathbf{A_v}$ and $\mathbf{A_c}$ are $n \times m$ matrices and $\mathbf{C}$ is an $m \times m$ diagonal matrix.

An expression for the determinant of **Y** can be derived with the aid of the Binet–Cauchy theorem. This states that if a matrix **P** is of size $n \times m$, and a matrix **Q** is of size $m \times n$, where $n < m$, then the determinant of the product **PQ** is given by

$$|\mathbf{PQ}| = \sum_{\substack{\text{all} \\ \text{majors}}} \left\{ \begin{array}{l} \text{products of corresponding } n \times n \\ \text{major determinants of } \mathbf{P} \text{ and } \mathbf{Q} \end{array} \right\} \tag{5.3.2}$$

Majors of **P** and **Q** are regarded as corresponding if the selection of columns from **P** matches the selection of rows from **Q**.

As an example of the operation of this theorem, consider the matrices:

$$\mathbf{P} = \begin{bmatrix} 1 & 0 & -2 \\ 3 & -1 & 1 \end{bmatrix} \qquad \mathbf{Q} = \begin{bmatrix} 2 & 1 \\ -1 & 4 \\ 0 & 3 \end{bmatrix}$$

The product of these matrices is given by

$$\mathbf{PQ} = \begin{bmatrix} 2 & -5 \\ 7 & 2 \end{bmatrix}$$

79

and

$$|\mathbf{PQ}| = 39$$

Evaluating the determinant using the Binet–Cauchy theorem gives

$$
|\mathbf{PQ}| = \begin{vmatrix} 1 & 0 \\ 3 & -1 \end{vmatrix} \begin{vmatrix} 2 & 1 \\ -1 & 4 \end{vmatrix}
$$

$$
+ \begin{vmatrix} 1 & -2 \\ 3 & 1 \end{vmatrix} \begin{vmatrix} 2 & 1 \\ 0 & 3 \end{vmatrix}
$$

$$
+ \begin{vmatrix} 0 & -2 \\ -1 & 1 \end{vmatrix} \begin{vmatrix} -1 & 4 \\ 0 & 3 \end{vmatrix}
$$

$$
= -9 + 42 + 6
$$

$$
= 39
$$

This agrees with the result obtained previously.

Returning now to the problem of evaluating the Y-matrix determinant, the Binet–Cauchy theorem can be applied to equation (5.3.1):

$$\Delta = |\mathbf{Y}| = |\mathbf{A_c} \ \mathbf{C} \ \mathbf{A_v^T}|$$

$$
= \sum_{\substack{\text{all} \\ \text{majors}}} \left\{ \begin{array}{l} \text{products of corresponding major} \\ \text{determinants of } \mathbf{A_c} \ \mathbf{C} \text{ and } \mathbf{A_v^T} \end{array} \right\} \tag{5.3.3}
$$

Now $\mathbf{C}$ is a diagonal matrix:

$$
\mathbf{C} = \begin{bmatrix} y_1 & & 0 \\ & y_2 & \\ 0 & & y_m \end{bmatrix} \tag{5.3.4}
$$

The product $\mathbf{A_c} \ \mathbf{C}$ therefore has the same structure as $\mathbf{A_c}$ except that each column j is multiplied by an admittance y_j.

In the previous section it was proved that a major determinant of the incidence matrix has a value of ± 1 if it corresponds to a tree of the network graph, otherwise it is zero. Suppose that a particular set of n branches is selected. If these branches form a tree of the current graph G_c then the major determinant of $\mathbf{A_c C}$ will equal the product of the branch admittances:

$$\pm y_{t1} y_{t2} y_{t3} \dots y_{tn} \tag{5.3.5}$$

where $t_1, t_2, t_3 \dots t_n$ are branches in the tree. Also, if the set of branches forms a tree of the voltage graph G_v then the major determinant of $\mathbf{A_v^T}$ will equal ± 1. The product of the major determinants will be zero unless the branches form spanning trees in both the voltage and current graphs.

Equation (5.3.3) can therefore be written in the form:

$$\Delta = \sum_W \pm \text{ (admittance products)} \tag{5.3.6}$$

where

$$W = \{\text{trees of } G_c\} \cap \{\text{trees of } G_v\}$$

W is the set of spanning trees that are common to both voltage and current graphs.

Passive networks have the property that their voltage and current graphs are identical, as are their voltage and current incidence matrices. This simplifies matters in two ways. Since the signs of the major determinants must be equal, their product must always be positive, and the trees of the voltage graph are also trees of the current graph. For a passive network with graph G, therefore, equation (5.3.6) reduces to

$$\Delta = \sum_W \text{ (admittance products)} \tag{5.3.7}$$

where

$$W = \{\text{trees of } G\}$$

To see how these results allow the determinant of the Y-matrix to be evaluated symbolically, consider the passive network shown in figure 5.1. This network has a Y-matrix given by

$$\mathbf{Y} = \begin{bmatrix} y_a & -y_a & 0 \\ -y_a & y_a + y_b + y_c & -y_b \\ 0 & -y_b & y_b + y_d + y_e \end{bmatrix}$$

Using the standard algebraic technique for evaluating determinants yields:

$$\begin{aligned} \Delta &= y_a\{(y_a + y_b + y_c)(y_b + y_d + y_e) - y_b^2\} + y_a\{-y_a(y_b + y_d + y_e)\} \\ &= y_a^2 y_b + y_a^2 y_d + y_a^2 y_e + y_a y_b^2 + y_a y_b y_d + y_a y_b y_e + y_a y_b y_c \\ &\quad + y_a y_c y_d + y_a y_c y_e - y_a y_b^2 - y_a^2 y_b - y_a^2 y_d - y_a^2 y_e \\ &= y_a y_b y_c + y_a y_b y_d + y_a y_b y_e + y_a y_c y_d + y_a y_c y_e \end{aligned}$$

Consider now the graph of this network, which is shown in figure 5.2. As has been demonstrated, this graph has five spanning trees, namely $\{abc\ abd\ abe\ acd\ ace\}$. The determinant of the Y-matrix is therefore given by the sum of these tree admittance products:

$$\Delta = y_a y_b y_c + y_a y_b y_d + y_a y_b y_e + y_a y_c y_d + y_a y_c y_e$$

As expected, this result is in agreement with the algebraically derived expression for Δ. The tree method, however, generated the terms in the expression directly, and without cancellation of terms.

5.4 Cofactors of the Y-Matrix

Network functions are given by the ratio of a cofactor of $\mathbf{Y}$ to the determinant of $\mathbf{Y}$, or by the ratio of two cofactors. In order to determine the network functions symbolically, therefore, an expression similar to equation (5.3.6) is required for the cofactors of the Y-matrix.

The cofactor Δ_{ij} of any matrix $\mathbf{Y}$ is given by

$$\Delta_{ij} = (-1)^{i+j} |\mathbf{Y}_{ij}| \tag{5.4.1}$$

where $\mathbf{Y}_{ij}$ is the matrix derived from $\mathbf{Y}$ by deleting the ith row and the jth column. Now, deleting the ith row of $\mathbf{Y}$ is equivalent to deleting the ith row of $\mathbf{A}_c$, and deleting the jth column of $\mathbf{Y}$ is equivalent to deleting the jth column of $\mathbf{A}_v^T$. But a column of $\mathbf{A}_v^T$ is a row of $\mathbf{A}_v$ so that

$$\Delta_{ij} = (-1)^{i+j} |\mathbf{A}_{ci} \ \mathbf{C} \ \mathbf{A}_{vj}^T| \tag{5.4.2}$$

where $\mathbf{A}_{ci}$ represents the matrix $\mathbf{A}_c$ with row i deleted, and $\mathbf{A}_{vj}$ represents the matrix $\mathbf{A}_v$ with row j deleted.

What operation on a network graph is equivalent to deleting a row from its incidence matrix? Rows of an incidence matrix correspond to network nodes. Coalescing a node with the reference node removes the corresponding row from the incidence matrix without affecting the branches or the other nodes. This is illustrated in figure 5.5. The graph G has an incidence matrix $\mathbf{A}$ given by

$$\mathbf{A} = \begin{matrix} & \begin{matrix} a & \ \ b & \ \ c & \ \ d & \ \ e \end{matrix} \\ \begin{matrix} 1 \\ 2 \\ 3 \end{matrix} & \begin{bmatrix} 1 & 0 & 0 & 0 & 0 \\ -1 & 1 & 1 & 0 & 0 \\ 0 & -1 & 0 & -1 & 1 \end{bmatrix} \end{matrix}$$

Graph G_1, which is derived from G by coalescing node 1 with the reference node 0, has an incidence matrix $\mathbf{A}_1$ given by

$$\mathbf{A}_1 = \begin{matrix} & \begin{matrix} a & \ \ b & \ \ c & \ \ d & \ \ e \end{matrix} \\ \begin{matrix} 2 \\ 3 \end{matrix} & \begin{bmatrix} -1 & 1 & 1 & 0 & 0 \\ 0 & -1 & 0 & -1 & 1 \end{bmatrix} \end{matrix}$$

As expected, $\mathbf{A}_1$ is equal to the original incidence matrix with row 1 deleted.

Figure 5.5
The effect of coalescing a node of the network graph with the reference node

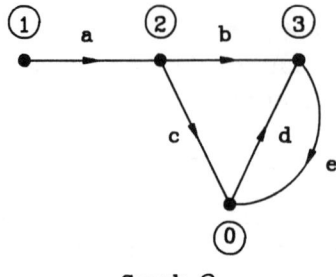

Graph G

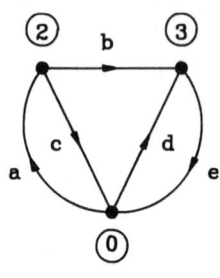

Graph G_1

It follows that if $\mathbf{A}$ is the incidence matrix of a graph G, then the non-zero major determinants of $\mathbf{A}_i$ (obtained from $\mathbf{A}$ by deleting row i) correspond to the spanning trees of the graph G_i (obtained from G by coalescing node i with the reference node). The same argument that was used to derive the expression for the determinant of $\mathbf{Y}$ can therefore be used to give the cofactors of $\mathbf{Y}$:

$$\Delta_{ij} = \sum_{W_{ij}} \pm \text{ (admittance products)} \qquad (5.4.3)$$

where

$$W_{ij} = \{\text{trees of } G_{ci}\} \cap \{\text{trees of } G_{vj}\}$$

A passive network can be represented by a single graph G. This leads to a simpler form for the cofactor of a passive network:

$$\Delta_{ij} = \sum_{W_{ij}} \text{ (admittance products)} \qquad (5.4.4)$$

where

$$W_{ij} = \{\text{trees of } G_i\} \cap \{\text{trees of } G_j\}$$

In the case of passive networks, all of the cofactor terms are positive. This is obvious for symmetrical cofactors, where

$$\Delta_{ii} = (-1)^{i+i} |\mathbf{A}_i \, \mathbf{C} \, \mathbf{A}_i^T| \qquad (5.4.5)$$
$$= |\mathbf{A}_i \, \mathbf{C} \, \mathbf{A}_i^T|$$

Since the corresponding major determinants of $\mathbf{A}_i\mathbf{C}$ and $\mathbf{A}_i^T$ have the same sign, the products must always be positive. Now the voltage transfer function of a network is given by

$$\frac{V_s}{V_t} = \frac{\Delta_{ts}}{\Delta_{tt}}$$

Δ_{tt} is positive so that the sign of the gain is the same as the sign of Δ_{ts}. But the d.c. gain of a passive network with a common reference node for input and output voltages is always positive. Since this fact is independent of the network topology and component values, all terms in Δ_{ts} must also be positive.

5.5 Symbolic Network Functions of Passive Networks

In the preceding sections, relationships were derived between the graph G of a passive network, and the determinant Δ and cofactors Δ_{ij} of the Y-

Figure 5.6
A twin-T filter

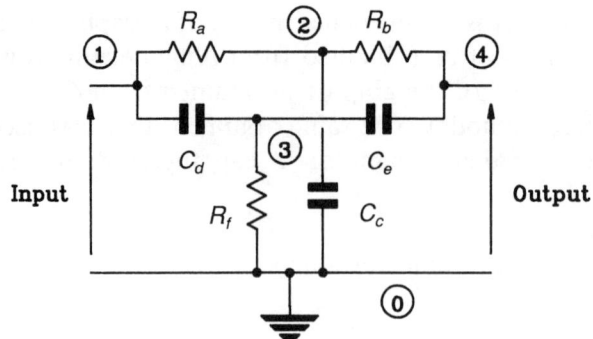

matrix:

$$\Delta = \sum_{W} (\text{admittance products})$$

$$\Delta_{ij} = \sum_{W_{ij}} (\text{admittance products}) \qquad (5.5.1)$$

where W = {trees of G}
and W_{ij} = {trees of G_i} $\cap$ {trees of G_j}

To see how the network functions can be derived using these formulae, consider the passive twin-T filter network shown in figure 5.6. The graph G of this network is shown in figure 5.7.

With a network graph of this simplicity it is possible to find the spanning trees by inspection. There are in fact twelve trees of G as shown in figure 5.8. These trees constitute the set W, and can be expressed more conveniently using the notation that a tree is represented by the product of its branches:

$$W = \{abcd\ abce\ abcf\ abdf\ abef\ acde$$
$$acef\ adef\ bcde\ bcdf\ bdef\ cdef\}$$

Figure 5.7
Network graph
of the twin-T
filter

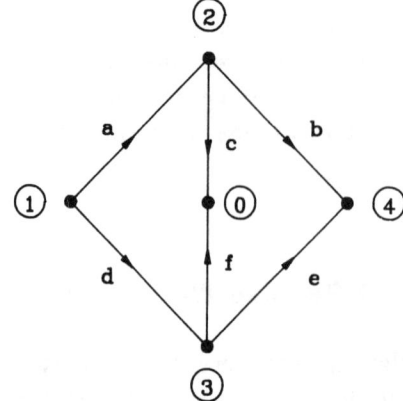

Figure 5.8
Spanning trees
of the network
graph

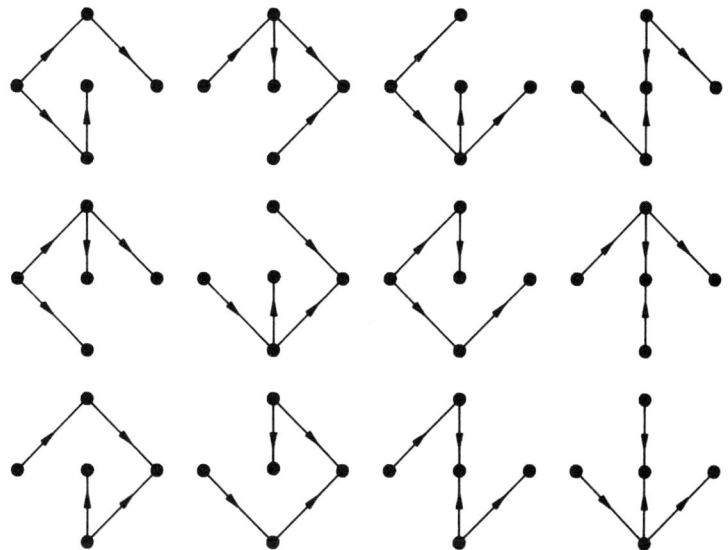

The voltage transfer function is given by the ratio of the cofactors Δ_{14} and Δ_{11}. These in turn are derived from the graphs G_1 and G_4, in which nodes 1 and 4 respectively of the network graph have been coalesced with the reference node, as illustrated in figure 5.9. Again, the spanning trees of these graphs can be found by inspection. As an example, the trees of G_1 are shown in figure 5.10.

Expressed algebraically the trees of G_1 and G_4 are given by

$$\{\text{trees of } G_1\} = \{abd \; abe \; abf \; ade \; aef \; bcd \\ bce \; bcf \; bde \; bef \; cde \; cef\}$$

$$\{\text{trees of } G_4\} = \{abd \; abe \; abf \; acd \; ace \; acf \\ ade \; adf \; bde \; bdf \; cde \; cdf\}$$

It is quite coincidental that there are the same number of trees of G and of

Figure 5.9
Graphs
obtained from
the network
graph by
coalescing
nodes 1 and 4
with the
reference node

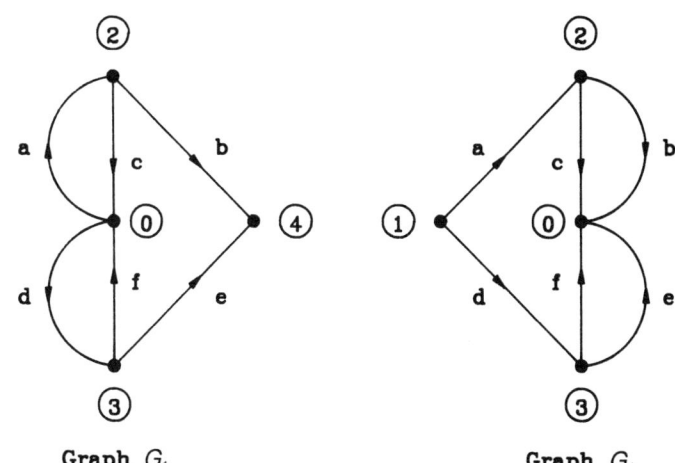

Graph G_1   **Graph G_4**

85

Figure 5.10
Spanning trees
of the graph G_1

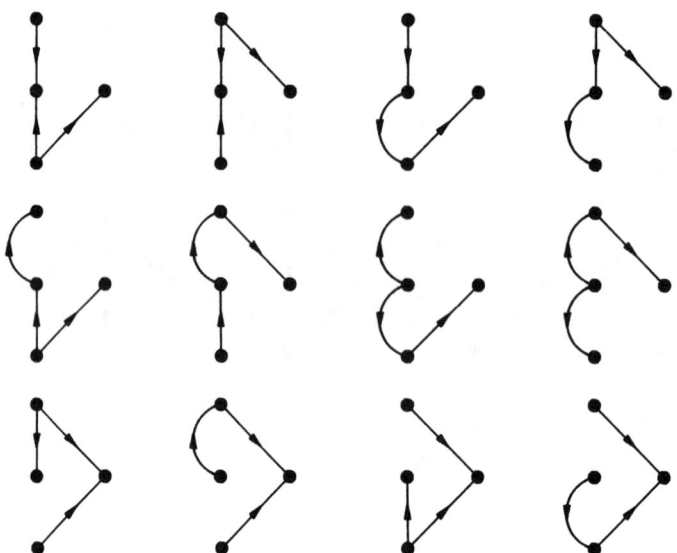

G_1; and G_1 and G_4 have the same number of trees because the network is symmetrical with respect to nodes 1 and 4.

Finally, the sets W_{11} and W_{14}, which are required for obtaining the cofactors, are given by

$$W_{11} = \{\text{trees of } G_1\} \cap \{\text{trees of } G_1\}$$
$$= \{\text{trees of } G_1\}$$
$$= \{abd \ abe \ abf \ ade \ aef \ bcd$$
$$bce \ bcf \ bde \ bef \ cde \ cef\}$$
$$W_{14} = \{\text{trees of } G_1\} \cap \{\text{trees of } G_4\}$$
$$= \{abd \ abe \ abf \ ade \ bde \ cde\}$$

The symbolic transfer function of the twin-T filter is therefore

$$\frac{V_4}{V_1} = \frac{\Delta_{14}}{\Delta_{11}}$$

$$= \frac{y_a y_b y_d + y_a y_b y_e + y_a y_b y_f + y_a y_d y_e + y_b y_d y_e + y_c y_d y_e}{\begin{array}{c} y_a y_b y_d + y_a y_b y_e + y_a y_b y_f + y_a y_d y_e + y_a y_e y_f + y_b y_c y_d \\ + y_b y_c y_e + y_b y_c y_f + y_b y_d y_e + y_b y_e y_f + y_c y_d y_e + y_c y_e y_f \end{array}}$$

Other symbolic network functions can also be obtained. For example, the input impedance of the twin-T filter is given by

$$\frac{V_1}{I_1} = \frac{\Delta_{11}}{\Delta}$$

$$= \frac{\begin{array}{c} y_a y_b y_d + y_a y_b y_e + y_a y_b y_f + y_a y_d y_e + y_a y_e y_f + y_b y_c y_d \\ + y_b y_c y_e + y_b y_c y_f + y_b y_d y_e + y_b y_e y_f + y_c y_d y_e + y_c y_e y_f \end{array}}{\begin{array}{c} y_a y_b y_c y_d + y_a y_b y_c y_e + y_a y_b y_c y_f + y_a y_b y_d y_f \\ + y_a y_b y_e y_f + y_a y_c y_d y_e + y_a y_c y_e y_f + y_a y_d y_e y_f \\ + y_b y_c y_d y_e + y_b y_c y_d y_f + y_b y_d y_e y_f + y_c y_d y_e y_f \end{array}}$$

With a network of the complexity of the twin-T filter it is reasonable to determine the trees of the various graphs by inspection. As the network complexity increases, however, this method rapidly becomes impracticable. A systematic spanning tree enumeration method, which can be expressed as a computer algorithm, is required if this symbolic analysis technique is to be of practicable use.

5.6 Spanning Tree Enumeration

A spanning tree enumeration algorithm must find all of the trees of a graph without duplication; it should also be as efficient as possible. At first sight the problem appears to be quite straightforward. A graph of r nodes will have trees which consist of $(r-1)$ branches. All of the possible combinations of $(r-1)$ branches are therefore generated, and those which do not form trees are discarded. Unfortunately, it can easily be seen that the number of combinations requiring to be tested increases very rapidly with the complexity of the graph and soon exceeds the capability of even the most powerful computers.

Consider a graph consisting of r nodes and m branches. The number of combinations of $(r-1)$ branches is given by

$$\binom{m}{r-1} = \frac{m!}{(r-1)!(m-r+1)!} \tag{5.6.1}$$

The graph shown in figure 5.11 contains 7 nodes and 15 branches. There are a total of 5005 combinations to be tested. Assuming that it takes 1 msec to determine whether a set of branches forms a tree, the complete tree enumeration procedure would occupy around 5 sec. This is normally an acceptable time to wait for an analysis.

The situation is quite different for a graph of twice this size, however. A graph of 14 nodes and 30 branches (which is the sort of complexity found commonly, for example, in active filter networks) has a total of around 10^8 combinations to be tested. Assuming, again, that it takes 1 msec to test each combination, the complete tree enumeration process will occupy 30 hours.

Figure 5.11
Graph of a passive ladder filter

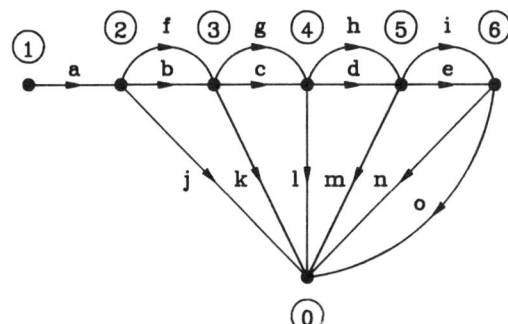

It is clear from the above discussion that generating all of the possible combinations of branches which might form trees is impracticable, except for fairly simple graphs. In fact, only a small proportion of the possible combinations of branches form trees. Surprisingly, it is easy to determine the number of spanning trees of a graph without actually enumerating them.

A spanning tree of a graph corresponds to a non-zero major determinant of the incidence matrix $\mathbf{A}$. Such major determinants have a value ± 1. Terms in the Binet–Cauchy expansion of $|\mathbf{A}\,\mathbf{A}^T|$ therefore have a value $+1$, and there is one term for each tree of the graph. Thus the value of $|\mathbf{A}\,\mathbf{A}^T|$ is simply equal to the number of trees. For the graph of figure 5.11:

$$|\mathbf{A}\,\mathbf{A}^T| = \begin{vmatrix} 1 & -1 & 0 & 0 & 0 & 0 \\ -1 & 4 & -2 & 0 & 0 & 0 \\ 0 & -2 & 5 & -2 & 0 & 0 \\ 0 & 0 & -2 & 5 & -2 & 0 \\ 0 & 0 & 0 & -2 & 5 & -2 \\ 0 & 0 & 0 & 0 & -2 & 4 \end{vmatrix}$$

$$= 512$$

For this particular network, therefore, only about 10% of the possible combinations of branches form trees; this proportion tends to become even smaller for more complex graphs.

It is clear from a brief inspection of figure 5.11 that a considerable saving in time can be achieved by not generating certain combinations of branches. For example, node 1 is connected to the rest of the graph by the single branch a. All trees of the graph must therefore contain branch a, and it is a waste of time generating combinations which do not contain a. Also, branches b and f form a loop and no tree can contain both b and f. A successful tree enumeration algorithm must recognize sets of branches which can never form part of a tree, and eliminate them at an early stage.

The spanning tree enumeration algorithm that will be described here is based on a simple formula for the trees of a graph G:

$$\begin{aligned} \{\text{trees of } G\} = \quad & \{\text{trees of a graph obtained from} \\ & \quad G \text{ by coalescing nodes } i,\, j\} \\ \times\ & \{\text{branches connecting } i,\, j\} \\ +\ & \{\text{trees of a graph obtained from} \\ & \quad G \text{ by removing branches between } i,\, j\} \end{aligned} \qquad (5.6.2)$$

This formula gives the trees of G in terms of the trees of two simpler graphs, one having less nodes, the other having less branches than G. It can be applied recursively to the original graph until, at some level of recursion, the graphs become trivial. (A graph is considered to be trivial when either it consists of a single node or it is not connected.) Figure 5.12 illustrates the

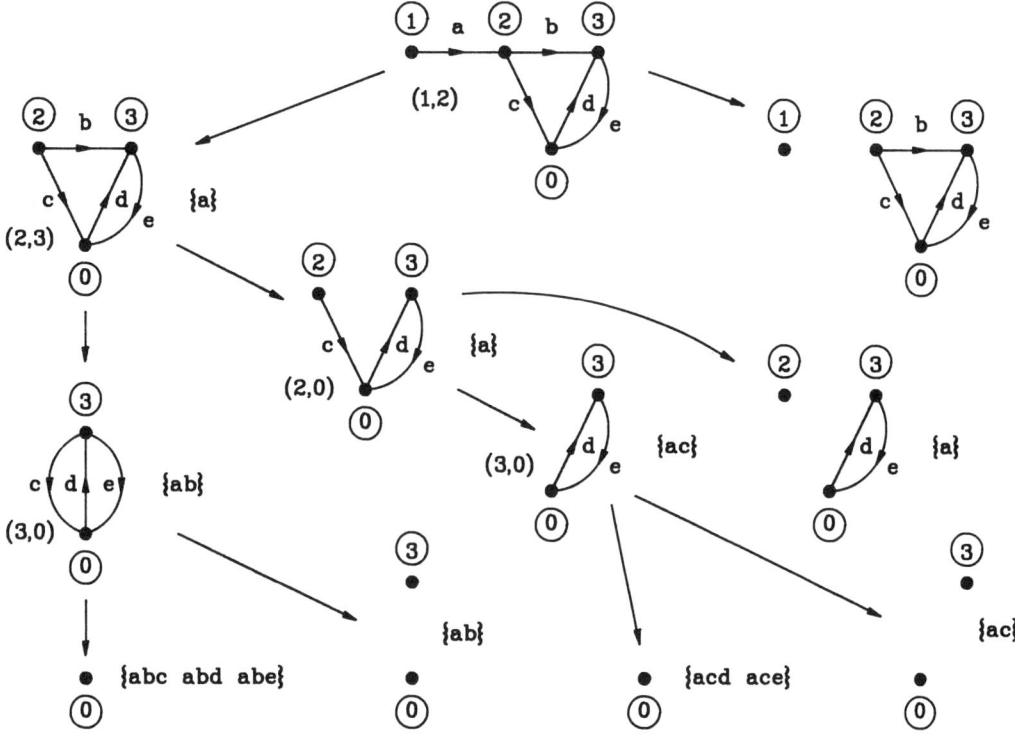

Figure 5.12 Spanning tree enumeration

operation of this formula when applied to a graph whose trees are {*abc abd abe acd ace*}.

At each stage the left-hand fork is the result of coalescing a pair of nodes, and the right-hand fork is the result of removing any branches between the nodes. Nodes 1 and 2 are chosen for the initial application of the formula. The initial right-hand fork is unconnected and cannot form a tree; no further reduction of this graph is therefore necessary. The initial left-hand fork leads eventually to two single-node graphs with associated branch products {*abc abd abe*} and {*acd ace*}.

Although the formula can be translated directly into any computer language that supports recursion, this does not lead to an efficient program. This is because each recursive call involves making a copy of the current graph to be passed as a parameter. A more efficient way of implementing the algorithm is to keep a single copy of the original graph, together with a tally of the coalesced nodes, and the eliminated branches.

Before the tree enumeration algorithm can be discussed in detail, a suitable computer representation of graphs must be established. Nodes of a graph are normally identified by positive integers, with node 0 being the reference node. The subrange of cardinals 0 ... 99 will be sufficient to represent the nodes of any network that is likely to be analyzed:

```
CONST
    maxno = 99;
TYPE
    node = [0..maxno];
```

Branches can be represented by their two terminating nodes t_1 and t_2, with the orientation defined to be away from t_1 and towards t_2:

```
TYPE
    branch = RECORD t1, t2: node END;
```

Finally, a graph can be represented by its branches; it is unnecessary also to specify the nodes since these are given by the branch terminations:

```
CONST
    maxbr = 80;
TYPE
    graph = ARRAY [1..maxbr] OF branch;
```

The procedure *tree* shown below finds all of the trees of the graph *G*; and *nno* and *nbr* are respectively the number of nodes and the number of branches in *G*:

```
TYPE
    brlist = ARRAY [1..maxbr] OF CARDINAL;

PROCEDURE tree(G: graph; nno, nbr: CARDINAL);
VAR
    sp, i: CARDINAL;
    stack: brlist;
    used: ARRAY [1..maxbr] OF BOOLEAN;
    intree: ARRAY node OF BOOLEAN;
```

```
BEGIN
    sp := 0;
    FOR i := 1 TO maxbr DO used[i] := FALSE END;
    FOR i := 0 TO maxno DO intree[i] := FALSE END;
    intree[G[1].t1] := TRUE;
    grow
END tree;
```

Starting from any node, a tree is grown by adding branches, one at a time, until all of the nodes are connected and the tree is complete. At each stage a branch is sought which connects the current sub-tree to a node outside the sub-tree. The key recursive procedure is *grow*, which is local to *tree*:

```
PROCEDURE grow;
VAR
    br: CARDINAL;
    nout: node;
BEGIN
    IF sp = nno-1 THEN foundtree(stack, sp)
    ELSIF findbr(br, nout) THEN
        push(br); used[br] := TRUE; intree[nout] := TRUE;
        grow;
        intree[nout] := FALSE; pull(br);
        grow;
        used[br] := FALSE
    END
END grow;
```

The procedure first checks to see whether a complete tree has been generated and, if so, executes *foundtree*. This procedure, which is not shown, might, for example, print out the trees or evaluate the corresponding admittance products. If the tree is not complete then *findbr* is called to search for a branch out of the current sub-tree. Two recursive calls to *grow* are then made, one with the new node effectively coalesced and the other with the new branch effectively removed.

91

The procedure *findbr*, which is local to *tree*, is shown below:

```
PROCEDURE findbr(VAR br: CARDINAL; VAR nout: node): BOOLEAN;
VAR
    nin: node;
BEGIN
    br := nbr;
    LOOP
        IF NOT used[br] THEN
            nout := G[br].t1; nin := G[br].t2;
            IF intree[nin] <> intree[nout] THEN
                IF intree[nout] THEN nout := nin END;
                RETURN TRUE
            END
        END;
        IF br = 1 THEN RETURN FALSE END;
        DEC(br)
    END
END findbr;
```

The procedures *push* and *pull* maintain a stack on which the branches in the current sub-tree are stored:

```
PROCEDURE push(k: CARDINAL);
BEGIN
    INC(sp); stack[sp] := k
END push;

PROCEDURE pull(VAR k: CARDINAL);
BEGIN
    k := stack[sp]; DEC(sp)
END pull;
```

On a standard IBM personal computer this routine took 2 sec to find the 512 trees of the graph shown in figure 5.11.

5.7 Common Spanning Tree Enumeration

When the determinants of active networks, or the unsymmetrical cofactors of passive networks, are evaluated, the admittance products correspond to the intersection of two sets of trees:

$$\Delta = \sum_W \pm \text{ (admittance products)}$$

$$\Delta_{ij} = \sum_{W_{ij}} \pm \text{ (admittance products)}$$

where $\quad W = \{\text{trees of } G_c\} \cap \{\text{trees of } G_v\}$

and $\quad W_{ij} = \{\text{trees of } G_{ci}\} \cap \{\text{trees of } G_{vj}\}$

In the previous section an algorithm was developed for finding the spanning trees of a single graph. Can this algorithm be used to find the spanning trees that are common to two graphs? One obvious approach is to use the algorithm to find the trees of each graph separately. The intersection set can then be generated by sorting the two sets of trees, and merging them to obtain their intersection. Unfortunately this approach makes enormous demands on storage to accommodate the two sets, each of which may consist of a large number of branch combinations.

An alternative method, which does not require any additional storage, is to generate the trees of the first graph, testing each of them to determine whether they are also trees of the second graph. Clearly this is rather inefficient because only a small proportion of the trees of one graph will also be trees of the other graph. However, it is quite adequate for use in analyzing networks of low to medium complexity.

One method for determining whether a set of branches forms a tree operates in a similar manner to the procedure *tree* described in the previous section. Starting from any node, a tree is grown by adding branches one at a time. If $(r - 1)$ branches can be incorporated (where r is the number of nodes) without generating any loops, then the branches form a tree. The procedure *istree* shown below returns a boolean value *TRUE* if the graph G is a tree. It makes use of a local procedure *findbr* which is identical to that described in the previous section.

```
PROCEDURE istree(G: graph; nno, nbr: CARDINAL): BOOLEAN;
VAR
    br, i: CARDINAL;
    nout: node;
    used: ARRAY [1..maxbr] OF BOOLEAN;
    intree: ARRAY node OF BOOLEAN;
```

```
BEGIN
    IF nbr <> nno-1 THEN RETURN FALSE END;
    FOR i := 1 TO maxbr DO used[i] := FALSE END;
    FOR i := 0 TO maxno DO intree[i] := FALSE END;
    intree[G[1].t1] := TRUE;
    FOR i := 1 TO nbr DO
        IF findbr(br, nout) THEN
            used[br] := TRUE; intree[nout] := TRUE
        ELSE
            RETURN FALSE
        END
    END;
    RETURN TRUE
END istree;
```

More sophisticated methods, which generate the common trees of two graphs directly, are available. Although these are more efficient than the method described above, they are also much more complicated and are outside the scope of this book.

5.8   Symbolic Network Functions of Active Networks

In the previous chapter it was shown that the topology of an active network can be represented by two graphs: the voltage graph G_v and the current graph G_c. For example, consider the active filter shown in figure 5.13. The network graphs are obtained in two stages. First, voltage and current graphs are drawn incorporating only the passive components. At this stage the graphs are identical, as shown in figure 5.14.

Then the nodes of the voltage graph corresponding to the operational amplifier inputs are coalesced, and the nodes of the current graph

Figure 5.13
A Sallen-key
active filter

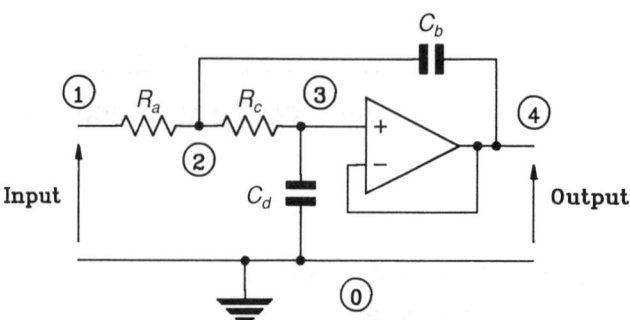

Figure 5.14
Graphs
incorporating
only passive
components

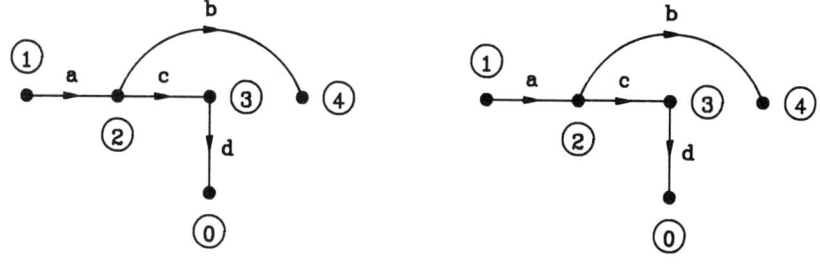

Voltage Graph G_v **Current Graph** G_c

corresponding to the operational amplifier outputs are coalesced. The resulting graphs, which are shown in figure 5.15, now represent the topology of the active network.

The voltage transfer function is given by the ratio of cofactors Δ_{14} and Δ_{11}, which in turn are derived from the graphs G_{c1}, G_{v4} and G_{v1}. These graphs are shown in figure 5.16. The spanning trees of these graphs are given by

$$\{\text{trees of } G_{v4}\} = \{ab \ ac\}$$
$$\{\text{trees of } G_{v1}\} = \{ab \ ac \ ad \ bd \ cd\}$$
$$\{\text{trees of } G_{c1}\} = \{ac \ ad \ bc \ bd \ cd\}$$

In order to evaluate the cofactors, the intersection sets W_{14} and W_{11} are

Figure 5.15
Voltage and
current graphs
of the filter

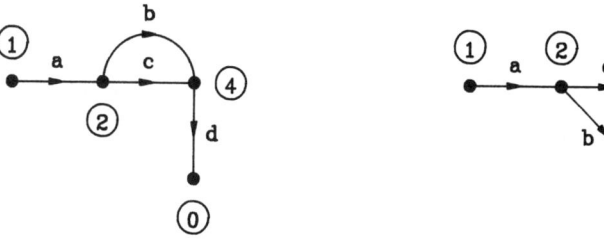

Voltage Graph G_v **Current Graph** G_c

Figure 5.16
Graphs required
for evaluation
of symbolic
transfer
function

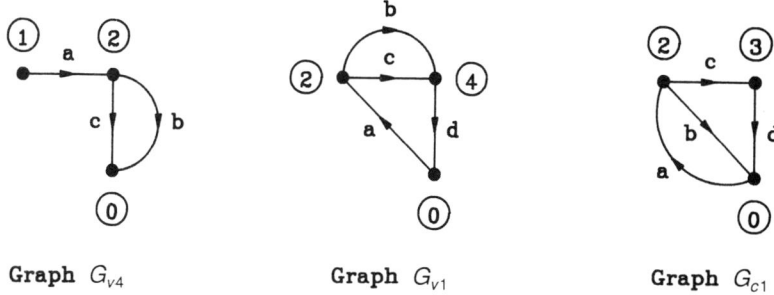

Graph G_{v4} **Graph** G_{v1} **Graph** G_{c1}

95

required:

$$W_{14} = \{\text{trees of } G_{c1}\} \cap \{\text{trees of } G_{v4}\}$$
$$= \{ac\}$$
$$W_{11} = \{\textit{trees of } G_{c1}\} \cap \{\text{trees of } G_{v1}\}$$
$$= \{ac \ ad \ bd \ cd\}$$

Finally, the symbolic transfer function of the active filter is given by

$$\frac{V_4}{V_1} = \frac{\Delta_{14}}{\Delta_{11}}$$

$$= \frac{\pm y_a y_c}{\pm y_a y_c \pm y_a y_d \pm y_b y_d \pm y_c y_d}$$

One problem remains, however. Unlike passive networks, where the signs of the admittance products are always positive, active networks have a sign uncertainty. In order to resolve this uncertainty it is necessary to return to the Binet–Cauchy expansions of the determinants and cofactors of the Y-matrix.

5.9 Signs of the Admittance Products

Unfortunately there is no simple way to obtain the signs of the admittance products directly from the network graphs. In the case of the determinant of the Y-matrix, the signs are given by

$$|\text{ major of } \mathbf{A}_c\|\text{ corresponding major of } \mathbf{A}_v^T|$$

Provided that the branches selected form a spanning tree of the network graph, then the corresponding major determinant of the incidence matrix has a value of ± 1. Cofactors of the Y-matrix require an additional sign correction factor of $(-1)^{i+j}$.

Major determinants can be evaluated by the Gaussian elimination procedure described in the previous chapter. For example, consider the network graph tree shown in figure 5.17. The major determinant of the

Figure 5.17
A network
graph tree

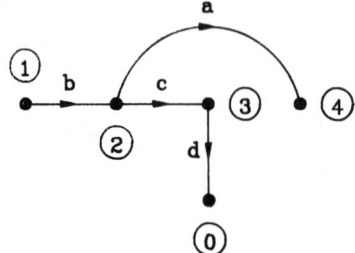

incidence matrix **A** corresponding to this tree is given by

$$| \text{major of } \mathbf{A} | = \begin{array}{c} \\ 1 \\ 2 \\ 3 \\ 4 \end{array} \begin{array}{cccc} a & b & c & d \\ \hline 0 & 1 & 0 & 0 \\ 1 & -1 & 1 & 0 \\ 0 & 0 & -1 & 1 \\ -1 & 0 & 0 & 0 \end{array}$$

The first stage in applying the Gaussian elimination procedure is to search the matrix for a row with a non-zero element in column a. (Since elements of the incidence matrix can only have the values 0 or ± 1, any non-zero element is suitable for use as the pivot.) Row 2 has this property, and is exchanged with row 1 with a consequent change in the sign of the determinant:

$$| \text{major of } \mathbf{A} | = - \begin{array}{c} \\ 2 \\ 1 \\ 3 \\ 4 \end{array} \begin{array}{cccc} a & b & c & d \\ \hline 1 & -1 & 1 & 0 \\ 0 & 1 & 0 & 0 \\ 0 & 0 & -1 & 1 \\ -1 & 0 & 0 & 0 \end{array}$$

Row 1 is now added to, or subtracted from, any other rows with a non-zero element in column 1. This has no effect on the sign of the determinant:

$$| \text{major of } \mathbf{A} | = - \begin{array}{c} \\ 2 \\ 1 \\ 3 \\ 4 \end{array} \begin{array}{cccc} a & b & c & d \\ \hline 1 & -1 & 1 & 0 \\ 0 & 1 & 0 & 0 \\ 0 & 0 & -1 & 1 \\ 0 & -1 & 1 & 0 \end{array}$$

This process is repeated for subsequent columns of the determinant, and leads to an upper triangular form for the matrix; all elements below the leading diagonal are zero:

$$| \text{major of } \mathbf{A} | = - \begin{array}{c} \\ 2 \\ 1 \\ 3 \\ 4 \end{array} \begin{array}{cccc} a & b & c & d \\ \hline 1 & -1 & 1 & 0 \\ 0 & 1 & 0 & 0 \\ 0 & 0 & -1 & 1 \\ 0 & 0 & 0 & 1 \end{array}$$

Now the determinant of such a triangular matrix is simply equal to the product of its diagonal elements. The final result is therefore:

$$| \text{major of } \mathbf{A} | = +1$$

This process of converting a matrix to triangular form will in general involve $n^3/3$ operations, where n is the size of the matrix. Incidence matrices, however, have a special structure which allows them to be converted to triangular form more efficiently.

Branches of the network graph terminate at two nodes. Since each column of the incidence matrix corresponds to one branch of the tree, there must be at least one, and at most two, non-zero elements in each column. Transposing the matrix, which of course does not alter the value of its determinant, leads to a matrix with at most two elements in each row. For example, for the tree shown in figure 5.17 the corresponding major determinant of the incidence matrix is given by

$$
|\,\text{major of }\mathbf{A}^T\,| = \begin{array}{c} \\ a \\ b \\ c \\ d \end{array} \begin{array}{|cccc|} 1 & 2 & 3 & 4 \\ \hline 0 & 1 & 0 & -1 \\ 1 & -1 & 0 & 0 \\ 0 & 1 & -1 & 0 \\ 0 & 0 & 1 & 0 \end{array}
$$

The elimination process is now simpler because instead of adding or subtracting a complete row it is only necessary to add or subtract, at most, two elements. The number of operations is therefore proportional to n^2 rather than n^3.

In order to take advantage of this simplification, the Gaussian elimination procedure must operate on the graph directly, rather than on its incidence matrix. A suitable procedure is shown below:

```
PROCEDURE det(G: graph; n: CARDINAL): REAL;
VAR
    i, j, k, p: CARDINAL;
    neg: BOOLEAN;
    t: branch;
BEGIN
    neg := FALSE;
    FOR i := 1 TO n DO
        p := i;
        WHILE (p <= n) AND (G[p].t1 <> i) AND (G[p].t2 <> i) DO
            INC(p)
        END;
        IF p > n THEN RETURN 0.0 END;
        IF p <> i THEN
            t := G[p]; G[p] := G[i]; G[i] := t;
            neg := NOT neg
        END;
```

```
        k := G[i].t1;
        IF k = i THEN k := G[i].t2 ELSE neg := NOT neg END;
        FOR j := i+1 TO n DO
            IF G[j].t1 = i THEN G[j].t1 := k
            ELSIF G[j].t2 = i THEN G[j].t2 := k END
        END
    END;
    IF neg THEN RETURN -1.0 ELSE RETURN 1.0 END
END det;
```

By using this procedure to evaluate the major determinants, together with the common spanning tree enumeration procedures, complete symbolic network functions can be generated. Consider the network shown in figure 5.18. It can be represented by the voltage and current graphs shown in figure 5.19.

The voltage transfer function of the filter is given by the ratio of the cofactors Δ_{14} and Δ_{11}. Since node 1 is the first node of the current graph (that is, the lowest numbered node excluding the reference node) and node 4 is the third node of the voltage graph, the respective cofactor sign corrections are $(-1)^{1+3}$ and $(-1)^{1+1}$. In order to evaluate the cofactors Δ_{14}

Figure 5.18
An all-pass filter

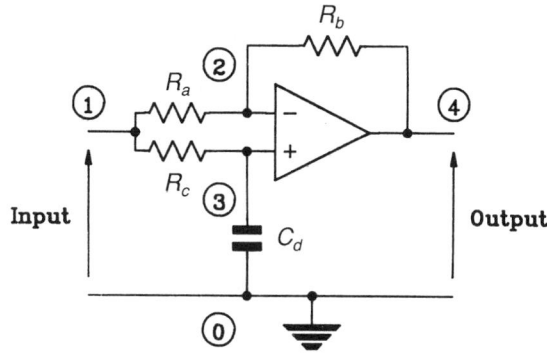

Input   Output

Figure 5.19
Voltage and current graphs of the all-pass filter

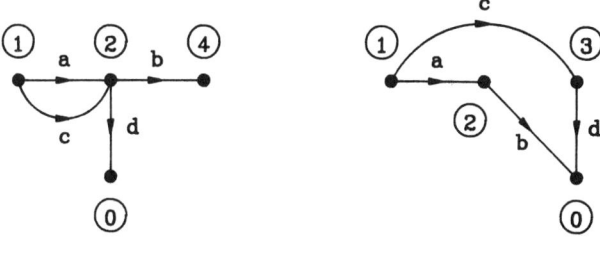

Voltage Graph G_v **Current Graph** G_c

Figure 5.20
Graphs required
for evaluation
of symbolic
transfer
function

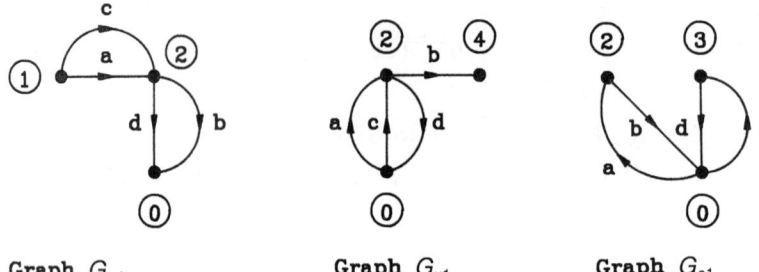

Graph G_{v4}   **Graph G_{v1}**   **Graph G_{c1}**

and Δ_{11} three graphs are required: G_{v4}, G_{v1} and G_{c1}. These graphs are shown in figure 5.20.

From the graphs the following sets of trees are obtained:

$$\{\text{trees of } G_{v4}\} = \{ab\ ad\ bc\ cd\}$$
$$\{\text{trees of } G_{v1}\} = \{ab\ bc\ bd\}$$
$$\{\text{trees of } G_{c1}\} = \{ac\ ad\ bc\ bd\}$$

The numerator of the transfer function is given by Δ_{14}, the terms of which correspond to set W_{14}:

$$W_{14} = \{\text{trees of } G_{c1}\} \cap \{\text{trees of } G_{v4}\}$$
$$= \{ad\ bc\}$$
$$\Delta_{14} = \pm y_a y_d \pm y_b y_c$$

Uncertainty in the signs can be resolved by evaluating the appropriate major determinants:

$$(\text{sign of } y_a y_d) = (-1)^{1+3}|\text{ major of } \mathbf{A_{c1}}||\text{ major of } \mathbf{A_{v4}}|$$

$$= (-1)^4 \quad \begin{array}{c} 2 \\ 3 \end{array}\begin{vmatrix} a & d \\ -1 & 0 \\ 0 & 1 \end{vmatrix} \quad \begin{array}{c} 1 \\ 2 \end{array}\begin{vmatrix} a & d \\ 1 & 0 \\ -1 & 1 \end{vmatrix}$$

$$= -1$$

$$(\text{sign of } y_b y_c) = (-1)^{1+3}\ |\text{ major of } \mathbf{A_{c1}}||\text{ major of } \mathbf{A_{v4}}|$$

$$= (-1)^4 \quad \begin{array}{c} 2 \\ 3 \end{array}\begin{vmatrix} b & c \\ 1 & 0 \\ 0 & -1 \end{vmatrix} \quad \begin{array}{c} 1 \\ 2 \end{array}\begin{vmatrix} b & c \\ 0 & 1 \\ 1 & -1 \end{vmatrix}$$

$$= +1$$

Thus

$$\Delta_{14} = y_b y_c - y_a y_d$$

By a similar procedure the cofactor Δ_{11} can be evaluated:

$$\Delta_{11} = y_b y_c + y_b y_d$$

Finally these results can be combined to give the symbolic transfer function of the all-pass filter:

$$\frac{V_4}{V_1} = \frac{\Delta_{14}}{\Delta_{11}} = \frac{y_b y_c - y_a y_d}{y_b y_c + y_b y_d}$$

5.10   Limitations of Symbolic Analysis

In the introduction to this chapter the many advantages of symbolic network analysis over numerical analysis were discussed. Since it is always possible to reduce a symbolic network function to numerical form, it might appear that there is never any need to perform numerical analysis. In fact this is not the case because symbolic network analysis has a serious drawback: the computational effort required to generate symbolic network functions increases extremely rapidly with network complexity.

In order to determine how critical this problem might be, a series of active ladder filters of increasing complexity were analyzed. For a network of 8 nodes the analysis time on a standard IBM personal computer was 2.2 sec. This rose to 28 sec for a 13-node network, to 660 sec for an 18-node network and to 17 000 sec for a 23-node network. It is clear from figure 5.21, where

Figure 5.21
Efficiency of network analysis methods

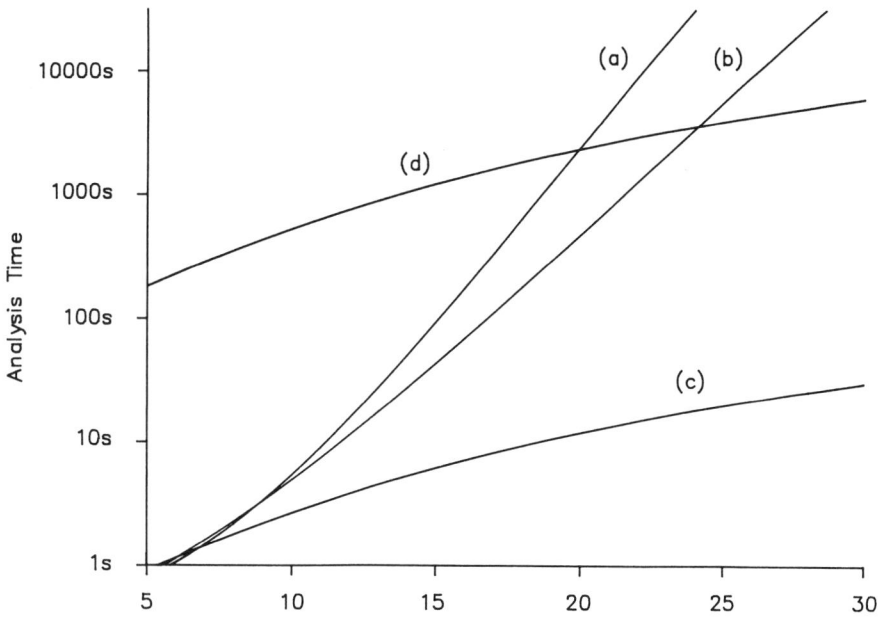

Network Complexity (Number of Nodes)

(a) Symbolic analysis

(b) Symbolic analysis (improved common tree algorithm)

(c) Numerical network analysis (1 frequency)

(d) Numerical network analysis (200 frequencies)

101

the logarithm of analysis time is plotted against network complexity, that the time for symbolic analysis increases exponentially with the number of nodes.

In most cases it would not be acceptable to wait for more than an hour for the result of a symbolic analysis, and this effectively places an upper limit of about 20 nodes on the complexity of networks that can be analyzed. Some improvement can be obtained by using a more sophisticated common spanning tree enumeration algorithm. As is shown in figure 5.21, analysis time is then reduced by a factor of 10, and the practical use of symbolic analysis is extended to networks of around 24 nodes.

Although the time for numerical analysis also increases with network complexity, it does so much less rapidly than is the case with symbolic analysis. Each numerical network function evaluation involves solving a set of n simultaneous equations (where n is the number of nodes excluding the reference node), and the time taken is proportional to n^3. Figure 5.21 shows the time for a single numerical analysis, and for 200 analyses (the number that would typically be required to plot a complete frequency response), of networks of various complexity.

If a single evaluation of some network function is required, then numerical analysis is always the most efficient technique. On the other hand, if 200 values are required, then it may be more efficient to perform a symbolic analysis and substitute appropriate component values and frequencies into the symbolic network function.

One obvious way of improving the efficiency of symbolic analysis is to partition the network, determine the network functions of each partition, and then combine them. For example, if a network of 24 nodes could be split into two networks of 12 nodes, then the analysis time would drop from 1 hour to 30 sec, a spectacular improvement. Unfortunately there is no generally applicable technique for partitioning networks. Certain classes of networks are, however, suitable for partitioning. In particular, cascade realizations of active filters, where the overall transfer function is equal to the product of the transfer functions of self-contained sections, can easily be partitioned.

In many cases a completely symbolic network function is not required and only one or two of the component values need to be kept in symbolic form. If this is so, then a modified form of numerical analysis can be used to obtain the semi-symbolic network function in a time proprotional to n^3. This method, which is known as parameter extraction, can be used for networks that are too complex to be analyzed fully symbolically.

5.11 Parameter Extraction

In general, large networks cannot be analyzed fully symbolically because of the amount of computation that is necessary. If the dependence of some network function on a single component value is all that is required, however, there is an alternative to full symbolic analysis.

Network functions are given by the ratio of a cofactor of the Y-matrix to the determinant of the Y-matrix, or by the ratio of two cofactors. Cofactors and determinants can be written as the sum of admittance products. The voltage transfer function of a network with input node t and output node s is therefore given by

$$\frac{V_s}{V_t} = \frac{\Delta_{ts}}{\Delta_{tt}} = \frac{\Sigma \pm (\text{admittance products})}{\Sigma \pm (\text{admittance products})} \tag{5.11.1}$$

Suppose that the dependence of the transfer function on the admittance α of a particular component is to be determined. The admittance products in the numerator and the denominator can be partitioned into those containing α, and those not containing α. Thus

$$\frac{V_s}{V_t} = \frac{A + \alpha B}{C + \alpha D} \tag{5.11.2}$$

Since the components may be reactive, the coefficients A, B, C, D are in general complex and frequency dependent.

A method will now be described for evaluating the dependence of the determinant of the Y-matrix on a particular admittance α. An equivalent method can be used for the cofactors of the Y-matrix. The determinant Δ of $\mathbf{Y}$ can be expressed in terms of the incidence matrices $\mathbf{A_c}$ and $\mathbf{A_v}$ and the branch admittance matrix $\mathbf{C}$ of the network:

$$\Delta = |\mathbf{Y}| = |\mathbf{A_c} \, \mathbf{C} \, \mathbf{A_v^T}| \tag{5.11.3}$$

Earlier in this chapter it was shown that Δ can be obtained from the network graphs G_v and G_c:

$$\Delta = \sum_W \pm (\text{admittance products}) \tag{5.11.4}$$

where $\quad W = \{\text{trees of } G_c\} \cap \{\text{trees of } G_v\}$

The set W can be partitioned into those trees which contain α and those trees which exclude α:

$$\begin{aligned} W = \; &\{\text{trees of } G_c \text{ exc. } \alpha\} \cap \{\text{trees of } G_v \text{ exc. } \alpha\} \\ &+ \{\text{trees of } G_c \text{ inc. } \alpha\} \cap \{\text{trees of } G_v \text{ inc. } \alpha\} \end{aligned} \tag{5.11.5}$$

Now the trees of a graph G which do not contain a particular branch are the same as the trees of a graph G' obtained from G by removing the branch. Also, the trees of a graph G which contain a particular branch are given by the product of this branch with the trees of a graph G'' obtained from G by coalescing the terminal nodes of the branch. This is illustrated in figure 5.22, in which an original graph is modified by alternatively removing, and coalescing, the branch d.

Figure 5.22
Effects of
removing and
coalescing a
branch

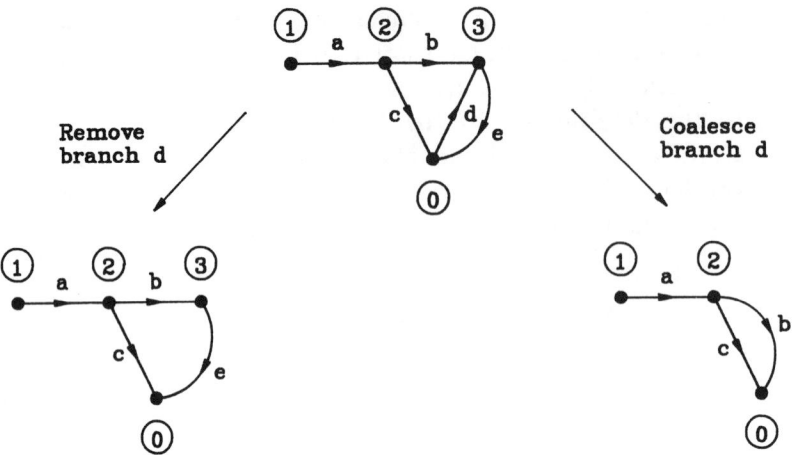

The trees of the graph with branch d removed are

{*abc abe ace*}

These are the trees of the original graph which do not contain d. The trees of the graph with branch d coalesced are

{*ab ac*}

Taking the product of these trees with the coalesced branch d gives those trees of the original graph which contain d:

{*abd acd*}

This method of partitioning can be applied to equation (5.11.4) to give

$$\Delta = \sum_{W'} \pm \text{ (admittance products)} \qquad (5.11.6)$$

$$+ \alpha \sum_{W''} \pm \text{ (admittance products)}$$

where $\quad W' = \{\text{trees of } G'_c\} \cap \{\text{trees of } G'_v\}$
and $\quad\quad W'' = \{\text{trees of } G''_c\} \cap \{\text{trees of } G''_v\}$

Finally, the determinant can be rewritten in matrix form:

$$\Delta = |\mathbf{Y}| = |\mathbf{Y}'| \pm \alpha |\mathbf{Y}''| \qquad (5.11.7)$$

In this equation $\mathbf{Y}'$ corresponds to the graphs G'_v and G'_c, and $\mathbf{Y}''$ corresponds to the graphs G''_v and G''_c.

Graphs G'_v and G'_c are obtained from G_v and G_c by removing the branch α. Thus $\mathbf{Y}'$ is the Y-matrix of the network excluding α. Graphs G''_v and G''_c are obtained from G_v and G_c by coalescing the branch α. Suppose that the branch terminates at nodes i, j in G_c and at k, l in G_v. The matrix $\mathbf{Y}''$ is derived from $\mathbf{Y}$ by the following sequence of operations. If either i or j is the reference node, then the row corresponding to the other node is deleted;

otherwise the row corresponding to node i is added to the row corresponding to node j, and is itself deleted. If either k or l is the reference node, then the column corresponding to the other node is deleted; otherwise the column corresponding to node k is added to the column corresponding to node l, and is itself deleted.

In effect these operations reduce the number of occurrences of α in the matrix to a single element, either $+\alpha$ or $-\alpha$, and then delete the row r and column c containing this element. If r and c correspond to the nodes i and k, or to the nodes j and l, then the deleted element Y_{rc} contains $+\alpha$; if r and c correspond to the nodes i and l, or to the nodes j and k then the deleted element contains $-\alpha$. This is one factor contributing to the sign in equation (5.11.7). A further sign correction of $(-1)^{r+c}$ is necessary because of the position of the term $+\alpha$, or $-\alpha$, in the matrix $\mathbf{Y}$. Combining these sign corrections gives the following expression for the determinant:

$$\Delta = |\mathbf{Y}| = |\mathbf{Y}'| + \alpha(-1)^{r+c+q}|\mathbf{Y}''| \tag{5.11.8}$$

where $\quad q = 1 \quad$ if $i = 0$ and $k \neq 0$, or $i \neq 0$ and $k = 0$;
otherwise $\quad q = 0$.

To see how parameter extraction works in practice, consider the network shown in figure 5.23. This represents a differential amplifier with inputs connected together in order to determine its common-mode gain.

The Y-matrix representing the passive components is given by

$$
\mathbf{Y} =
\begin{array}{c}
 \\
1 \\
2 \\
3 \\
4
\end{array}
\begin{array}{cccc}
\quad 1 \quad & \quad 2 \quad & \quad 3 \quad & \quad 4 \quad \\
\left[\begin{array}{cccc}
10^{-3} + \alpha & -10^{-3} & -\alpha & 0 \\
-10^{-3} & 10^{-3} + 10^{-5} & 0 & -10^{-5} \\
-\alpha & 0 & 10^{-5} + \alpha & 0 \\
0 & -10^{-5} & 0 & 10^{-5}
\end{array}\right]
\end{array}
$$

Now the operational amplifier can be embedded in the network by deleting the row corresponding to node 4, and combining the columns corresponding to nodes 2 and 3. During these operations it is necessary to keep a track of

Figure 5.23
A differential
amplifier with
inputs
connected
together

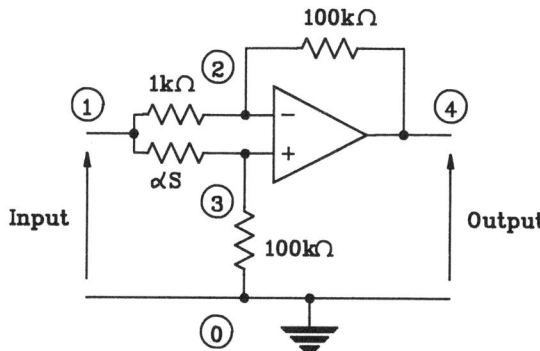

100kΩ

1kΩ

Input

αS

100kΩ

Output

105

the terminal nodes i, k, j, l of the admittance α.

$$
\mathbf{Y} = \begin{array}{c} \\ 1 \\ 2 \\ 3 \end{array}
\begin{array}{c}
 1 3 4 \\
\left[\begin{array}{ccc}
10^{-3} + \alpha & -10^{-3} - \alpha & 0 \\
-10^{-3} & 10^{-3} + 10^{-5} & -10^{-5} \\
-\alpha & 10^{-5} + \alpha & 0
\end{array} \right]
\end{array}
$$

$i = 1, \; j = 3, \; k = 1, \; l = 3$

The voltage transfer ratio of the network is given by the cofactors Δ_{14} and Δ_{11} of the Y-matrix:

$$
\frac{V_4}{V_1} = \frac{\Delta_{14}}{\Delta_{11}}
$$

Consider the first numerator cofactor Δ_{14}. This is obtained from the Y-matrix by deleting the row corresponding to node 1 and the column corresponding to node 4:

$$
\Delta_{14} = (-1)^{1+3} \begin{array}{c} 2 \\ 3 \end{array}
\begin{array}{c}
 1 3 \\
\left| \begin{array}{cc}
-10^{-3} & 10^{-3} + 10^{-5} \\
-\alpha & 10^{-5} + \alpha
\end{array} \right|
\end{array}
$$

$i = 0, \; j = 3, \; k = 1, \; l = 3$

Applying equation (5.11.8) to this cofactor allows the parameter α to be extracted:

$$
\Delta_{14} = \left| \begin{array}{cc}
-10^{-3} & 10^{-3} + 10^{-5} \\
0 & 10^{-5}
\end{array} \right| + \alpha (-1)^{2+1+1} | 10^{-5} |
$$

$$
= -10^{-8} + 10^{-5} \alpha
$$

The denominator cofactor Δ_{11} is given by

$$
\Delta_{11} = (-1)^{1+1} \begin{array}{c} 2 \\ 3 \end{array}
\begin{array}{c}
 3 4 \\
\left| \begin{array}{cc}
10^{-3} + 10^{+5} & -10^{-5} \\
10^{-5} + \alpha & 0
\end{array} \right|
\end{array}
$$

$i = 0, \; j = 3, \; k = 0, \; l = 3$

Again, applying equation (5.11.8) allows the parameter α to be extracted:

$$
\Delta_{11} = \left| \begin{array}{cc}
10^{-3} + 10^{-5} & -10^{-5} \\
10^{-5} & 0
\end{array} \right| + \alpha (-1)^{2+1+0} | -10^{-5} |
$$

$$
= 10^{-10} + 10^{-5} \alpha
$$

Finally, the expressions for the cofactors can be combined to give the voltage transfer ratio:

$$
\frac{V_4}{V_1} = \frac{-10^{-8} + 10^{-5} \alpha}{10^{-10} + 10^{-5} \alpha}
$$

Evaluation of a network function in parameter extracted form involves the calculation of four determinants of complex matrices. The procedure for evaluating determinants is very similar to the Gaussian elimination technique for solving simultaneous equations that was described in the previous chapter. Two important properties of determinants make this possible: an interchange of any two rows simply changes the sign of the determinant, and the addition or subtraction of a proportion of one row to another does not affect the value of the determinant.

The first step is to search column 1 for the best pivot. If this pivot is not in row 1 then a row interchange with row 1 is performed, and a change of sign is registered. Row 1 is now added, or subtracted, in various proportions to the remaining rows in order to eliminate elements in column 1. This procedure is repeated for each column and the result is an upper triangular matrix with all elements below the leading diagonal being zero. The value of the determinant is then equal to the product of the diagonal elements with a sign inversion if an odd number of row interchanges were performed.

Evaluation of the determinant of a real matrix involves $n^3/3$ multiplications. In the previous chapter, a method for solving complex simultaneous equations by turning them into an equivalent set of real equations was described. Unfortunately there is no similar method for dealing with complex determinants. Each multiplication in the evaluation procedure must therefore be performed on complex quantities and will involve four real multiplications; a total of $4n^3/3$ multiplications are required to evaluate a complex determinant.

Since the time taken to evaluate a complex determinant is comparable with the time taken to solve an equivalent set of complex simultaneous equations, parameter extraction should take four times as long as numerical analysis. For large networks it is far more efficient than fully symbolic analysis.

5.12 Summary

Symbolic network analysis allows network functions to be obtained in the form of rational functions of jω, where the coefficients are symbolic expressions. Symbolic network functions can be converted to numerical form by substitution of component values; substitution of the frequency reduces numerical network functions to complex numbers.

Symbolic network analysis has a number of advantages over numerical analysis. It does not suffer from the accuracy problems associated with numerical analysis because terms are cancelled symbolically. The effects of varying particular component values can be derived directly from symbolic network functions; this is important in sensitivity analysis and network optimization. Another use of symbolic analysis is in time-domain response calculations. Inverse Laplace transformation of numerical network functions allows the impulse and step responses of a network to be determined. Finally, if a network function is available in symbolic form then the component values of the network can be chosen to give a desired frequency response.

Network functions such as input impedance and voltage transfer function can be obtained from the determinant and cofactors of the Y-matrix; these in turn can be deduced from the network graphs. In the case of passive networks, whose topology is defined by a single graph, the determinant of the Y-matrix is given by the admittance products corresponding to spanning trees of the network graph. Cofactors of the Y-matrix are given by the admittance products corresponding to common spanning trees of two graphs.

Spanning trees of a graph can be enumerated with the aid of a simple formula which gives the trees in terms of the trees of two simpler graphs, one having less nodes, the other having less branches, than the original graph. This formula is applied recursively until, at some level of recursion, the graphs become trivial. It is possible to find the common spanning trees of two graphs using this algorithm: the trees of one graph are enumerated, and each is tested to see whether it also forms a tree of the other graph.

Active networks, whose topology is defined by a pair of graphs, can be analyzed in a similar manner. There is, however, an additional complication. Unlike passive networks, for which the admittance products are always positive, active networks have admittance products which may be of either sign. This sign uncertainty can be resolved by evaluating the major determinants of the incidence matrices. Gaussian elimination is normally used for this purpose. If advantage is taken of the special structure of incidence matrices, then an $n \times n$ determinant can be evaluated using n^2 operations.

Symbolic network analysis is limited in its applications by the fact that the computational effort increases exponentially with network complexity. For large networks, therefore, symbolic analysis is less efficient than numerical analysis.

Parameter extraction is an alternative to full symbolic analysis when the dependence of some network function on a single component value is required. It involves the evaluation of four complex determinants and the number of operations is proportional to n^3, where n is the number of network nodes. Unlike symbolic analysis, however, parameter extraction must be performed at each frequency of interest.

5.13 Problems

1 Draw the network graph of the high-pass doubly-terminated ladder filter shown in figure 5.24. Find the spanning trees of the graphs obtained by:

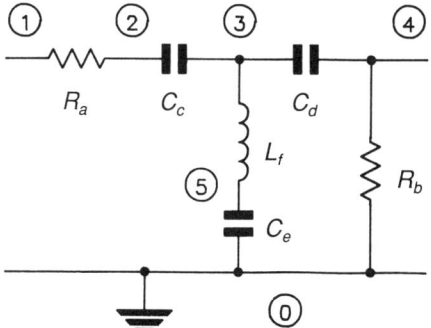

Figure 5.24
A doubly-
terminated
ladder filter

    (*a*) coalescing nodes 1 and 0 of the network graph;
    (*b*) coalescing nodes 4 and 0 of the network graph.

Hence derive symbolic expressions for the cofactors Δ_{11} and Δ_{14} of the Y-matrix. Use these results to obtain the transfer function V_4/V_1 in symbolic form.

2   Figure 5.25 shows a resistive attenuation network. Draw the network graph of this attenuator. Find the spanning trees of the network graph, and of the graph obtained by coalescing nodes 1 and 0 of the network graph. Hence derive symbolic expressions for the determinant Δ and the cofactor Δ_{11} of the Y-matrix. Use these results to obtain the input impedance V_1/I_1 in symbolic form.

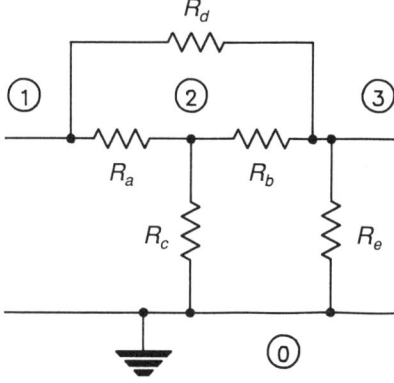

Figure 5.25
A resistive
attenuator

3   The network shown in figure 5.26 represents the small-signal equivalent circuit of a bipolar transistor amplifier with parallel feedback. Draw the voltage and current graphs of this network. Find the spanning trees of the graphs obtained by

    (*a*) coalescing nodes 1 and 0 of the current graph;
    (*b*) coalescing nodes 1 and 0 of the voltage graph;
    (*c*) coalescing nodes 3 and 0 of the voltage graph.

Hence derive symbolic expressions for the cofactors Δ_{11} and Δ_{13} of the Y-

Figure 5.26
Equivalent
circuit of a
transistor
amplifier

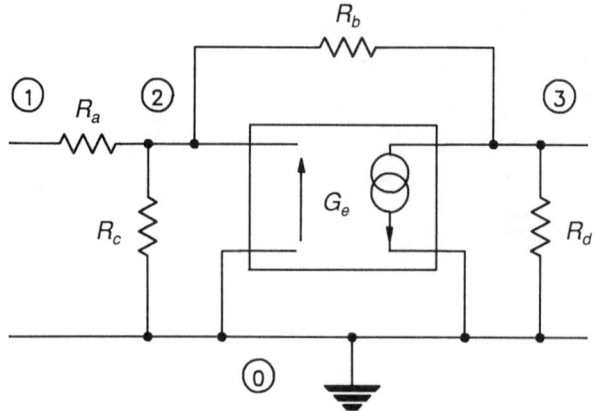

matrix, resolving sign uncertainties by evaluating the major determinants. Use these results to obtain the transfer function V_3/V_1 in symbolic form.

4 Draw the voltage and current graphs of the active filter network shown in figure 5.27. Find the spanning trees of the graphs obtained by

(*a*) coalescing nodes 1 and 0 of the current graph;
(*b*) coalescing nodes 1 and 0 of the voltage graph;
(*c*) coalescing nodes 4 and 0 of the voltage graph.

Figure 5.27
A Rauch filter

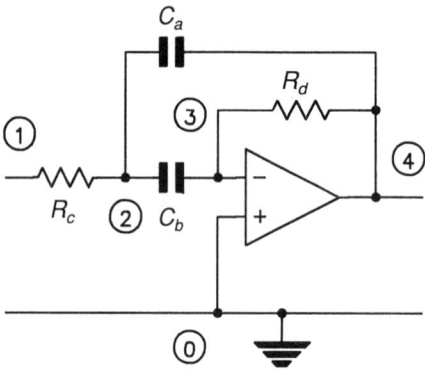

Figure 5.28
A resistive
attenuator

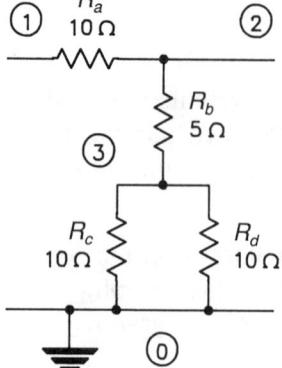

Hence derive symbolic expressions for the cofactors Δ_{11} and Δ_{14} of the Y-matrix, resolving sign uncertainties by evaluating the major determinants. Use these results to obtain the transfer function V_4/V_1 in symbolic form.

5 Figure 5.28 shows a resistive attentuator network. Use the method of parameter extraction to determine the dependence of the transfer function V_2/V_1 on the value of the resistor R_d.

6 Time-domain Analysis of Linear Networks

6.1 Introduction

In some applications only the frequency-domain response of an electronic system is of interest. For example, an anti-aliasing filter for speech signals would normally be specified in the frequency domain, and the time-domain response would be of secondary importance. At the other extreme, some electronic systems are specified by their time-domain response only. Typical of such systems are matched filters which are used to extract pulse signals from noise. (The impulse response of a matched filter is the time reverse of the signal it is designed to detect.) More common than either of these extremes are systems in which both time- and frequency-domain performances are of importance.

The starting point for time-domain analysis is usually the transfer function $H(s)$. If the time-domain response of an existing network is required, then its transfer function can be obtained by symbolic analysis. If, on the other hand, a network is to be designed with some particular time- and frequency-domain properties, then the design procedure normally consists of two independent stages. In the first stage the transfer function which meets the specification is determined; the second stage consists of implementing the required transfer function by an appropriate electronic network.

Three methods for determining the time-domain response of linear electronic systems will be considered. The inverse Laplace transform approach is straightforward and efficient, but can suffer from problems of poor numerical accuracy in some circumstances. It is also limited to a few basic input functions. Numerical convolution can be used to determine the response of a linear system to any input, provided that the impulse response of the system is known. Finally, the state variable method is a general technique for representing linear systems in the time domain. It can be used to determine the response to a variety of inputs and does not suffer from the accuracy problems of the inverse Laplace transform approach.

6.2 The Inverse Laplace Transform Approach

The output $y(t)$ of a linear system with transfer function $H(s)$ and input $x(t)$ is given by

$$y(t) = \mathscr{L}^{-1}Y(s) \tag{6.2.1}$$

where $\quad Y(s) = H(s)X(s)$

and $\quad X(s) = \mathscr{L}x(t)$

It is in principle possible to determine $y(t)$ by performing numerically an inverse Laplace transform on $Y(s)$:

$$y(t) = \frac{1}{2\pi j} \int_{\sigma - j\infty}^{\sigma + j\infty} Y(s)e^{st}\, ds \tag{6.2.2}$$

Integration is performed along a path on the complex plane parallel to the $j\omega$ axis. Numerical evaluation of this integral is both complicated and time-consuming, and the integral must be evaluated for each value of t. Plotting the response of a system might well involve one hundred integrals.

Fortunately, provided that the input transform $X(s)$ is a rational function, it is possible to manipulate $Y(s)$ algebraically into a sum of simpler functions whose inverse Laplace transforms are known. The time-domain response $y(t)$ is then obtained by adding together these individual inverse transforms.

Three input functions whose Laplace transforms are rational functions are the unit impulse $\delta(t)$, the unit step $u(t)$, and the exponential decay:

$$\begin{aligned}
\mathscr{L}\,\delta(t) &= 1 \\
\mathscr{L}u(t) &= 1/s \\
\mathscr{L}e^{\alpha t} &= 1/(s - \alpha)
\end{aligned} \tag{6.2.3}$$

Since $H(s)$ is itself a rational function (provided that it represents a system containing no distributed elements), the output transform $Y(s) = H(s)X(s)$ will also be a rational function:

$$Y(s) = \frac{c_0 + c_1 s + c_2 s^2 + \cdots + c_n s^n}{d_0 + d_1 s + d_2 s^2 + \cdots + d_n s^n} \tag{6.2.4}$$

For example, the procedure *makestep* shown below generates the output transform $Y(s)$ of a linear system $H(s)$ with a unit-step function input by, in effect, multiplying $H(s)$ by $1/s$:

```
PROCEDURE makestep(H: rational; VAR Y: rational);
VAR
    i: CARDINAL;
```

```
BEGIN
    Y.order := H.order+1;
    FOR i := 0 TO H.order DO
        Y.a[i] := H.a[i];
        Y.b[i+1] := H.b[i]
    END;
    Y.a[Y.order] := 0.0; Y.b[0] := 0.0
END makestep;
```

Using one of the methods described in chapter 3 the denominator polynomial of $Y(s)$ is factorized to give

$$Y(s) = \frac{c_0 + c_1 s + c_2 s^2 + \cdots + c_n s^n}{d_n(s - p_1)(s - p_2)\ldots(s - p_n)} \tag{6.2.5}$$

If there are no coincident poles, then a partial fraction expansion of this expression can be performed giving $Y(s)$ as a sum of first-order terms:

$$Y(s) = Q + \frac{k_1}{s - p_1} + \frac{k_2}{s - p_2} + \cdots + \frac{k_n}{s - p_n}$$

$$= Q + \sum_{i=1}^{n} \frac{k_i}{s - p_i} \tag{6.2.6}$$

where $Q = c_n/d_n$.

The coefficients k_i are known as the residues of their respective poles and can be evaluated by means of the so-called cover-up rule. Suppose that the residue k_j of a pole p_j is to be determined. Both sides of equation (6.2.6) are multiplied by $(s - p_j)$ to give

$$(s - p_j)Y(s) = (s - p_j)Q + \frac{(s - p_j)k_1}{s - p_1} + \cdots + k_j + \cdots + \frac{(s - p_j)k_n}{s - p_n} \tag{6.2.7}$$

If the value of s is set to p_j, then all of the terms on the right-hand side of this equation become zero except k_j. The residue k_j is therefore given by

$$k_j = [(s - p_j)Y(s)]_{s=p_j} \tag{6.2.8}$$

The product $(s - p_j)Y(s)$ is not evaluated at $s = p_j$ by calculating the product of $(p_j - p_j)$ and $Y(p_j)$ because this would, of course, lead to an indeterminate result. Instead $Y(s)$ is calculated at $s = p_j$ omitting the factor $(s - p_j)$ in the denominator, hence the name 'cover-up rule'.

Consider the rational function shown below (which for simplicity has real poles):

$$Y(s) = \frac{s^3 - s - 8}{2s^3 - 12s^2 + 22s - 12}$$

The first step is to normalize the function by dividing numerator and denominator by the denominator coefficient of s^3:

$$Y(s) = \frac{0 \cdot 5s^3 - 0 \cdot 5s - 4}{s^3 - 6s^2 + 11s - 6}$$

Extracting the constant term (which is now equal to the numerator coefficient of s^3) gives

$$Y(s) = 0 \cdot 5 + \frac{3s^2 - 6s - 1}{s^3 - 6s^2 + 11s - 6}$$

The denominator can now be factorized. Since it is being assumed that there are no coincident poles, the Newton–Raphson method is quite suitable for this purpose.

$$Y(s) = 0 \cdot 5 + \frac{3s^2 - 6s - 1}{(s - 1)(s - 2)(s - 3)}$$

This expression can be written in partial fraction form, where the residues k_1, k_2, k_3 are yet to be determined:

$$Y(s) = 0 \cdot 5 + \frac{k_1}{s - 1} + \frac{k_2}{s - 2} + \frac{k_3}{s - 3}$$

Each of the residues can now be found by covering up the appropriate factor in the denominator. For example, the residue k_1 is given by

$$k_1 = \left[\frac{3s^2 - 6s - 1}{(s - 2)(s - 3)} \right]_{s=1}$$
$$= -4/2$$
$$= -2$$

A similar procedure gives the other residues:

$$k_2 = 1 \quad \text{and} \quad k_3 = 4$$

and the complete partial fraction expansion of $Y(s)$ becomes

$$Y(s) = 0 \cdot 5 - \frac{2}{s - 1} + \frac{1}{s - 2} + \frac{4}{s - 3}$$

A procedure *expand* is given below which reduces a rational function Y to partial fraction form. The result is returned as a constant q and two complex arrays p, k containing the poles and their residues.

```
PROCEDURE expand(Y: rational; VAR q: REAL; VAR p, k: complexpoly);
VAR
    i, j, m: CARDINAL;
    t: REAL;
    f, g, x: complex;
    c: complexpoly;
```

```
BEGIN
    WITH Y DO
        m := order;
        q := a[m]/b[m];
        FOR i := 0 TO m DO
            c[i].re := (a[i]-q*b[i])/b[m]; c[i].im := 0.0;
            p[i].re := b[i]/b[m]; p[i].im := 0.0

        END

    END;
    newton(p, m);
    FOR j := 1 TO m DO
        evf(c, m, p[j], f);
        g.re := 1.0; g.im := 0.0;
        FOR i := 1 TO m DO
            IF i <> j THEN
                x.re := p[j].re-p[i].re;
                x.im := p[j].im-p[i].im;
                t := x.re*g.re-x.im*g.im;
                g.im := x.re*g.im+x.im*g.re;
                g.re := t

            END

        END;
        t := g.re*g.re+g.im*g.im;
        k[j].re := (f.re*g.re+f.im*g.im)/t;
        k[j].im := (f.im*g.re-f.re*g.im)/t

    END
END expand;
```

This procedure makes use of two other routines which are described in detail in chapter 3. The procedure *newton*, which finds the zeros of a polynomial equation, is used to determine the poles of Y and the procedure *evf* evaluates a polynomial function.

Once the output transform $Y(s)$ has been obtained in partial fraction form, an inverse Laplace transform can be performed on the individual terms. Using the Laplace transforms given in equations (6.2.3) the output function $y(t)$ for $t \geqslant 0$ can be written as

$$y(t) = \mathscr{L}^{-1} Y(s)$$

$$= \mathscr{L}^{-1} Q + \sum_{i=1}^{n} \mathscr{L}^{-1} \frac{k_i}{s - p_i} \qquad (6.2.9)$$

$$= Q\delta(t) + \sum_{i=1}^{n} k_i e^{p_i t}$$

Real poles, where $p_i = \sigma_i + 0j$, give rise to simple exponential decays:

$$k_i e^{\sigma_i t} \qquad (6.2.10)$$

Complex poles, where $p_i = \sigma_i + j\omega_i$, lead to exponentially decaying sinusoids:

$$k_i e^{\sigma_i t} e^{j\omega_i t} = k_i e^{\sigma_i t} \{\cos \omega_i t + j \sin \omega_i t\} \qquad (6.2.11)$$

Since the denominator coefficients of $Y(s)$ are real, complex poles must occur in conjugate pairs. The output function therefore contains terms of the form:

$$k_i e^{\sigma_i t} e^{j\omega_i t} + k_j e^{\sigma_i t} e^{-j\omega_i t}$$
$$= e^{\sigma_i t} \{(k_i + k_j)\cos \omega_i t + j(k_i - k_j)\sin \omega_i t\} \qquad (6.2.12)$$

Of course, the output function $y(t)$ is real. It follows that the residues k_i, k_j of a conjugate pair of poles are also conjugate and the corresponding time-domain response simplifies to

$$2 \, |k_i| \, e^{\sigma_i t} \cos(\omega_i t + \phi_i) \qquad (6.2.13)$$

Consider the network shown in figure 6.1, for which the unit-step response is to be determined. Analysis of this network shows that the transfer function $H(s)$ is

$$H(s) = \frac{1 + (1 \cdot 67 \times 10^{-8} s^2)}{1 + (5 \cdot 00 \times 10^{-4} s) + (1 \cdot 00 \times 10^{-7} s^2) + (8 \cdot 33 \times 10^{-12} s^3)}$$

The Laplace transform of a unit-step function input is $X(s) = 1/s$ so that the output transform $Y(s)$ is given by

$$Y(s) = \frac{1 + (1 \cdot 67 \times 10^{-8} s^2)}{s + (5 \cdot 00 \times 10^{-4} s^2) + (1 \cdot 00 \times 10^{-7} s^3) + (8 \cdot 33 \times 10^{-12} s^4)}$$

Figure 6.1
A third-order
low-pass filter

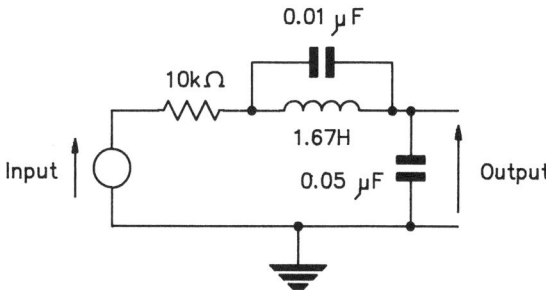

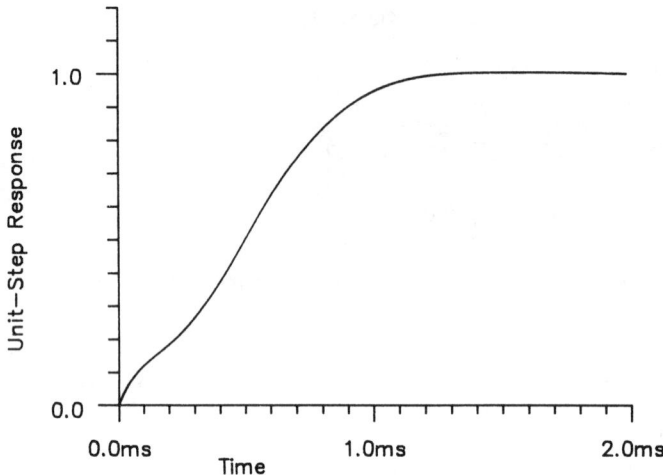

Figure 6.2
Step response of the third-order filter

This output transform can be expanded in partial fraction form:

$$Y(s) = \frac{1}{s} - \frac{2 \cdot 652}{s + 4644} + \frac{0 \cdot 826 + 0 \cdot 604j}{s + 3678 - 3509j} + \frac{0 \cdot 826 - 0 \cdot 604j}{s + 3678 + 3509j}$$

Finally an inverse transform can be applied to each of the terms to give the step response $y(t)$ for $t \geqslant 0$. For $t < 0$ the response is, of course, zero.

$$y(t) = 1 - 2 \cdot 652 e^{-4644t}$$
$$+ e^{-3678t} \{1 \cdot 652 \cos(3509t) - 1 \cdot 209 \sin(3509t)\}$$
$$= 1 - 2 \cdot 652 e^{-4644t} + 2 \cdot 047 e^{-3678t} \cos(3509t + 0.632)$$

Once the response has been obtained in this form, it can be evaluated rapidly by substitution of appropriate values of t. Whenever possible the response should be displayed graphically, as shown in figure 6.2, rather than as a table of corresponding values of y and t.

A procedure *ilp* is given below which performs an inverse Laplace transform on a partial fraction expansion defined by the constant q, and the complex arrays p, k containing the poles and their residues. The procedure returns the value of the inverse transform at a time t.

```
PROCEDURE ilp(t, q: REAL; p, k: complexpoly; n: CARDINAL): REAL;
VAR
    st, wt, y: REAL;
    i: CARDINAL;
BEGIN
    y := 0.0;
    IF t = 0.0 THEN
        IF q > 0.0 THEN y := infinity
        ELSIF q < 0.0 THEN y := -infinity END
    END;
```

```
    FOR i := 1 TO n DO
        st := t*p[i].re; wt := t*p[i].im;
        y := y + exp(st)*(k[i].re*cos(wt)-k[i].im*sin(wt))
    END;
    RETURN y
END ilp;
```

It is clear that the partial fraction expansion method for performing inverse Laplace transforms is a simple and efficient technique for the time-domain analysis of linear networks. Why then is it necessary to consider other analysis techniques? The main objection to partial fraction expansion is that it becomes numerically unstable if two poles approach one another. To understand the reason for this, consider the output transform $Y(s)$ shown below:

$$Y(s) = \frac{1}{s^2 + 2s + 0 \cdot 999999}$$

This transform represents the impulse response of a perfectly well-behaved system. The poles of $Y(s)$ are situated at $-1 \cdot 001$ and $-0 \cdot 999$ and the partial fraction expansion is therefore given by

$$Y(s) = \frac{k_1}{s + 1 \cdot 001} + \frac{k_2}{s + 0 \cdot 999}$$

Evaluating the residues using the cover-up rule gives

$$k_1 = \frac{1}{-1 \cdot 001 + 0 \cdot 999} = -500$$

$$k_2 = \frac{1}{-0 \cdot 999 + 1 \cdot 001} = 500$$

In both cases the subtraction of quantities around 1.0 gives rise to denominators of magnitude $0 \cdot 002$, leading to a loss of three decimal digits of accuracy in the residues. As the poles approach one another more closely the problem becomes increasingly severe. There is worse to come, however. Performing inverse Laplace transforms on the partial fractions gives the time-domain response $y(t)$:

$$y(t) = -500e^{-1 \cdot 001t} + 500e^{-0 \cdot 999t}$$

Close to $t = 0$ the response is the difference of two nearly equal quantities. For example, substituting $t = 0 \cdot 1$ sec in the above expression gives

$$y(0 \cdot 1) = -500e^{-0 \cdot 1001} + 500e^{-0 \cdot 0999}$$
$$= -452 \cdot 373 + 452 \cdot 463$$
$$= 0 \cdot 090$$

Another four significant figures have been lost in this subtraction leading to a cumulative loss of accuracy of seven decimal digits.

If a number of poles are known to be exactly coincident, then a partial fraction expansion can be performed which avoids these numerical difficulties. In practice, the numerical routines used to locate the poles will not generate identical values for pairs of poles, and even if it were possible to detect coincident poles the problem of close but non-coincident poles still remains.

The partial fraction method for performing an inverse Laplace transform cannot be considered to be a general method for time-domain analysis. When the poles of the transfer function are known to be well-separated, however, it is a valuable technique which is simple and more efficient than alternative methods.

6.3 Numerical Convolution

Time-domain analysis by inverse Laplace transformation is only suitable for use with certain input functions. Only if the transform of the input is a rational function is it possible to perform a partial fraction expansion of the output transform. In practice this restricts the technique to impulse and step function inputs. More complex input functions can be dealt with by numerical convolution.

In chapter 2 it was shown that the output $y(t)$ of a linear system is given by the convolution of the system impulse response $h(t)$ with the input $x(t)$:

$$y(t) = \int_{-\infty}^{\infty} x(\tau)h(t - \tau) \, d\tau \tag{6.3.1}$$

As was demonstrated in the previous section, the impulse response can be obtained in analytical form by inverse Laplace transformation of $H(s)$. Calculating the response to a general input $x(t)$ therefore involves performing a numerical integration each time the output is to be evaluated. Fortunately, although the impulse response of the system extends from $t = 0$ to $t = \infty$, the input function $x(t)$ will normally be bounded in time. This means that the limits of integration in the convolution need not cover an infinite range. If the input function is zero outside the range $t = t_1$ to $t = t_2$ then equation (6.3.1) becomes

$$y(t) = \int_{t_1}^{t_2} x(\tau)h(t - \tau) \, d\tau \tag{6.3.2}$$

A further reduction in the integration range becomes possible if it is noted that the impulse response is zero for $t < 0$:

$$y(t) = \int_{t_1}^{\min(t,t_2)} x(\tau)h(t - \tau) \, d\tau \tag{6.3.3}$$

Suppose, for example, that the response of the third-order filter shown in figure 6.1 to a Gaussian input pulse, of width $0\cdot5$ ms and centred on $t = 2$ ms, is required. The input pulse is effectively zero outside the range $t = 0$ to $t = 4$ ms:

$$x(t) = \exp\left\{ -\frac{(t - 2 \times 10^{-3})^2}{2(0\cdot5 \times 10^{-3})^2} \right\} \qquad \text{for } 0 \leqslant t \leqslant 4 \text{ ms}$$

$$= 0 \qquad \text{for } t < 0 \text{ and } t > 4 \text{ ms}$$

Partial fraction expansion of the transfer function, followed by inverse Laplace transformation gives the impulse response:

$$h(t) = 12317 e^{-4644t} + 10404 e^{-3678t} \cos(3509t + 3\cdot011)$$

This is illustrated in figure 6.3.

A numerical integration procedure, such as the procedure *adaptive* described in chapter 7, can now be used to calculate the response to the Gaussian input pulse:

```
t := 0.0;
WHILE t <= tmax DO
     y := adaptive(convolve, 0.0, min(t, 4.0E-3), 1.0E-3);
     WriteReal(t, 12); WriteString('    ');
     WriteReal(y, 12); WriteLn;
     t := t+dt
END;
```

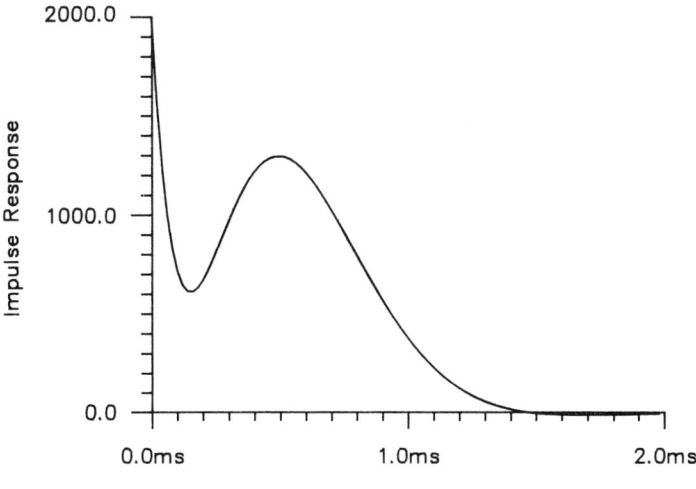

Figure 6.3
Impulse
response of the
third-order filter

121

The procedure *convolve* returns the product of the input function and the impulse response of the system:

```
PROCEDURE gaussian(t: REAL): REAL;
CONST
    t0 = 2.0E-3;
    dt = 0.5E-3;
BEGIN
    RETURN exp(-(t-t0)*(t-t0)/(2.0*dt*dt))
END gaussian;

PROCEDURE convolve(tau: REAL): REAL;
BEGIN
    RETURN gaussian(tau)*ilp(t-tau, q, p, k, H.order)
END convolve;
```

This method was used to calculate the response of the third-order filter to the Gaussian pulse at intervals of $0 \cdot 05$ ms from $t = 0$ to $t = 5$ ms and the result is shown in figure 6.4. The output pulse is Gaussian in appearance, but has been somewhat broadened, and has been delayed by approximately $0 \cdot 5$ ms. On a standard IBM personal computer the time taken to evaluate each point on this response was around 4 sec and plotting the complete response took about 7 minutes.

Figure 6.4
Response of the filter to a Gaussian pulse

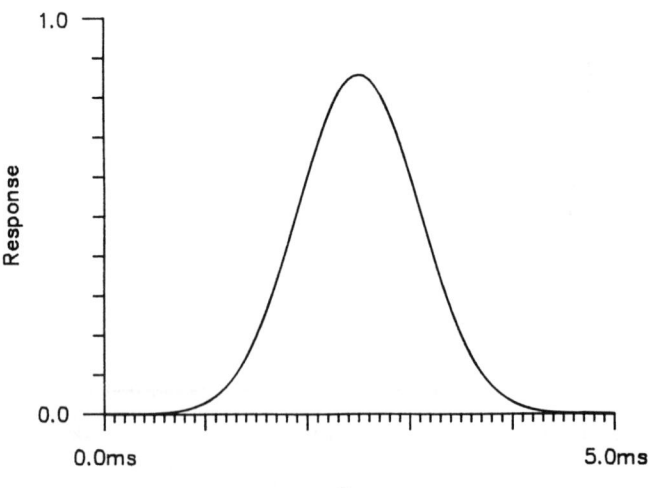

122

6.4 State Variable Representation

The time-domain behaviour of linear electrical networks containing no distributed elements can be represented by a set of first-order differential equations. These differential equations are known as the state variable equations and are usually written in vector–matrix form:

$$\dot{\mathbf{x}}(t) = \mathbf{A}\,\mathbf{x}(t) + \mathbf{B}u(t)$$
$$y(t) = \mathbf{C}^T\mathbf{x}(t) + Du(t)$$

(6.4.1)

In these equations $u(t)$ is the input to the network and $y(t)$ is the output; $\mathbf{x}(t)$ is a vector of internal, or state, variables. The matrix $\mathbf{A}$, the vectors $\mathbf{B}$, $\mathbf{C}$ and the scalar D are not related directly to any of the network matrices discussed in previous chapters, but define completely the dynamical behaviour of the network.

Time-domain analysis of linear networks using the state variable method involves two separate stages. In the first stage the coefficients $\mathbf{A}$, $\mathbf{B}$, $\mathbf{C}$ and D are determined, either directly from the network, or from its transfer function $H(s)$. Then the state equations are solved by some numerical procedure. The state variable approach has the advantage that it provides useful insights into the physical behaviour of networks and it can readily be extended to deal with non-linear networks.

The state vector $\mathbf{x}(t)$ consists of a set of independent variables whose values are sufficient to determine the state of a network, that is to determine all the branch voltages and currents. Not all of these branch voltages and currents are required to determine the state, however, since some of them can be calculated from the others. Although the state of a system is unique, its representation in terms of state variables is not. There are, in fact, an infinite number of state representations of any system. State variables are associated with energy storage elements (capacitors and inductors) in the network. To see why this is so, consider the simple resistor–capacitor network shown in figure 6.5.

The behaviour of this network is described by the equations:

$$i(t) \;\; = C\,\frac{\mathrm{d}}{\mathrm{d}t}\,v_c(t)$$
$$v_r(t) = Ri(t)$$
$$v_i(t) \;\; = v_r(t) + v_c(t)$$

(6.4.2)

Figure 6.5
A resistor–
capacitor filter

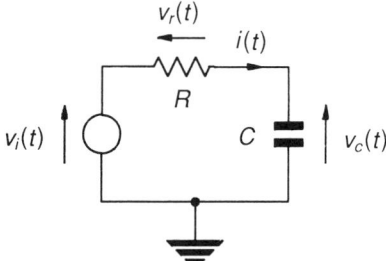

Eliminating the variables $i(t)$ and $v_r(t)$ leads to a simple equation relating the capacitor voltage to the input voltage:

$$\frac{d}{dt} v_c(t) = -\frac{1}{RC} v_c(t) + \frac{1}{RC} v_i(t) \qquad (6.4.3)$$

This differential equation can be solved to give the voltage v_c at a time T:

$$v_c(T) = e^{-T/RC} v_c(0) + \frac{1}{RC} \int_0^T v_i(t) e^{(t-T)/RC} \, dt \qquad (6.4.4)$$

All that is required to calculate $v_c(T)$ for any $T \geqslant 0$ is a knowledge of the initial value $v_c(0)$ and the input $v_i(t)$ between $t = 0$ and $t = T$; the variation of $v_i(t)$ prior to $t = 0$ is irrelevant. The other branch variables $i(t)$ and $v_r(t)$ can be calculated from $v_c(t)$ and $v_i(t)$, so that $v_c(t)$ completely defines the state of the network at a time t and can be taken to be the state variable.

This choice of state variable is not the only one possible, however. An alternative state variable might be $i(t)$, since $v_c(t)$ can be derived from $i(t)$ by

$$v_c(t) = v_i(t) - Ri(t) \qquad (6.4.5)$$

This illustrates the fact that the choice of state variables is somewhat arbitrary. In general, if $\mathbf{Q}$ is a square non-singular matrix, then a new set of state variables $\mathbf{z}$ can be defined by

$$\mathbf{x}(t) = \mathbf{Q} \, \mathbf{z}(t) \qquad (6.4.6)$$

Substituting this expression into the state equations (6.4.1) gives

$$\begin{aligned} \dot{\mathbf{z}}(t) &= \mathbf{Q}^{-1}\mathbf{A} \, \mathbf{Q} \, \mathbf{z}(t) + \mathbf{Q}^{-1}\mathbf{B}u(t) \\ y(t) &= \mathbf{C}^T\mathbf{Q} \, \mathbf{z}(t) + Du(t) \end{aligned} \qquad (6.4.7)$$

These equations constitute a new state variable representation of the system; the input–output relationship defined by these equations is, however, identical to that defined by the original equations. In practice the choice of state variables depends on whether the network itself, or only its transfer function, is available.

In the Modula-2 procedures that follow, variables of type *statevar* will be used to store the state variable representations of networks, where:

```
TYPE
    statevar = RECORD A: matrix;
                      B, C: vector;
                      D: REAL;
                      order: index
               END;
```

6.5 The Physical State Variables

If the starting point for time-domain analysis is the network itself then the state variables can be taken to be the physical state variables, that is the capacitor voltages and the inductor currents. One advantage of such a choice is that it simplifies the initial conditions since it can usually be assumed that the input is zero for $t < 0$. Capacitor voltages and inductor currents are therefore zero at $t = 0$.

In the case of canonic networks (a network is canonic if the total number of capacitors and inductors is equal to its order), all of the capacitor voltages and inductor currents are state variables. Consider the third-order filter network shown in figure 6.6. The voltages and currents indicated are functions of time, although their dependence is not explicitly stated. Analysis of the network gives the following equations which define the relationship between the input voltage v_{in} and the output voltage v_{out}:

$$(v_{in} - v_c)/R_a = i_e + C_c \frac{dv_c}{dt}$$

$$v_c - v_d = L_e \frac{di_e}{dt}$$

$$i_e = C_d \frac{dv_d}{dt} + v_d/R_b$$

$$v_{out} = v_d$$

These equations can be rearranged into the standard state variable form:

$$\begin{bmatrix} \dot{v}_c \\ \dot{i}_e \\ \dot{v}_d \end{bmatrix} = \begin{bmatrix} -1/R_a C_c & -1/C_c & 0 \\ 1/L_e & 0 & -1/L_e \\ 0 & 1/C_d & -1/R_b C_d \end{bmatrix} \begin{bmatrix} v_c \\ i_e \\ v_d \end{bmatrix} + \begin{bmatrix} 1/R_a C_c \\ 0 \\ 0 \end{bmatrix} v_{in}$$

$$v_{out} = \begin{bmatrix} 0 & 0 & 1 \end{bmatrix} \begin{bmatrix} v_c \\ i_e \\ v_d \end{bmatrix} + 0 v_{in}$$

Figure 6.6
A third-order
low-pass
Butterworth
filter

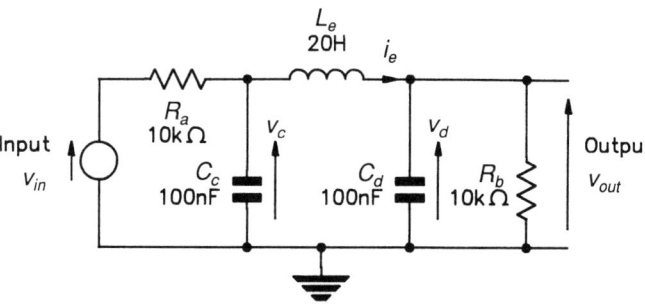

125

Figure 6.7
A third-order
elliptic filter

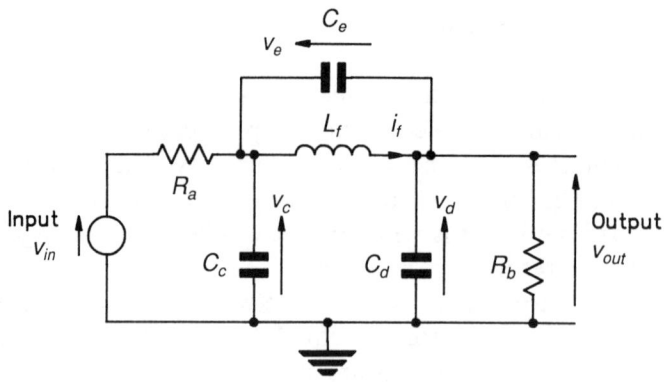

Finally the component values can be substituted to give

$$
\begin{bmatrix} \dot{v}_c \\ \dot{i}_e \\ \dot{v}_d \end{bmatrix} = \begin{bmatrix} -1 \times 10^3 & -1 \times 10^7 & 0 \\ 5 \times 10^{-2} & 0 & -5 \times 10^{-2} \\ 0 & 1 \times 10^7 & -1 \times 10^3 \end{bmatrix} \begin{bmatrix} v_c \\ i_e \\ v_d \end{bmatrix} + \begin{bmatrix} 1 \times 10^3 \\ 0 \\ 0 \end{bmatrix} v_{in}
$$

$$
v_{out} = \begin{bmatrix} 0 & 0 & 1 \end{bmatrix} \begin{bmatrix} v_c \\ i_e \\ v_d \end{bmatrix} + 0 v_{in}
$$

In this example all of the capacitor voltages and inductor currents are independent, but this is not always the case, as can be seen from a brief inspection of the network shown in figure 6.7. Any one of the capacitor voltages can be expressed in terms of the other two, for example:

$$
v_c = v_d + v_e
$$

Only two of the capacitor voltages, say v_c and v_d, together with the inductor current i_f, are therefore necessary to specify the state of the network.

In general the choice of an independent set of physical state variables may be by no means obvious. Formal methods do exist for selecting state variables, but they will not be described here because they are not necessary for the purpose of calculating the time-domain behaviour of networks. An alternative route to the state variable representation is via the transfer function. Any network can be analyzed symbolically using the techniques described in chapter 5. The transfer function can then be converted to state variable form, thus avoiding the difficulty of selecting physical state variables.

6.6   State Variable Representation from the Transfer Function

The state variable representation of a network can be derived from its transfer function, although the resulting state variables will not in general correspond directly with the physical state variables. For any given transfer function there are an infinite number of possible state variable representations and the representation that will be derived here is the so-called companion form.

It will be assumed that the transfer function is a rational function of s, so that the input and output transforms $U(s)$ and $Y(s)$ are related by

$$H(s) = \frac{Y(s)}{U(s)}$$

$$= \frac{a_0 + a_1 s + a_2 s^2 + \cdots + a_{n-1} s^{n-1} + a_n s^n}{b_0 + b_1 s + b_2 s^2 + \cdots + b_{n-1} s^{n-1} + b_n s^n} \qquad (6.6.1)$$

This can be written in the form:

$$(b_0 + b_1 s + b_2 s^2 + \cdots + b_{n-1} s^{n-1} + b_n s^n) Y(s)$$
$$= (a_0 + a_1 s + a_2 s^2 + \cdots + a_{n-1} s^{n-1} + a_n s^n) U(s) \qquad (6.6.2)$$

At this point it is convenient to introduce a new variable $Z(s)$ defined by

$$(a_0 + a_1 s + a_2 s^2 + \cdots + a_{n-1} s^{n-1} + a_n s^n) Z(s) = Y(s) \qquad (6.6.3)$$

Equation (6.6.2) then becomes

$$(b_0 + b_1 s + b_2 s^2 + \cdots + b_{n-1} s^{n-1} + b_n s^n) Z(s) = U(s) \qquad (6.6.4)$$

Now it is necessary to shift into the time domain with the aid of the following property of Laplace transforms:

$$\mathscr{L} \frac{\mathrm{d}}{\mathrm{d}t} f(t) = s\mathscr{L}f(t) - f(0-) \qquad (6.6.5)$$

Assuming zero initial conditions gives

$$\mathscr{L} \frac{\mathrm{d}}{\mathrm{d}t} f(t) = sF(s) \qquad (6.6.6)$$

where $F(s)$ is the Laplace transform of $f(t)$. From this it follows that

$$\mathscr{L} \frac{\mathrm{d}^n}{\mathrm{d}t^n} f(t) = s^n F(s) \qquad (6.6.7)$$

This result can be used to perform an inverse Laplace transformation on equation (6.6.4):

$$\left\{ b_0 + b_1 \frac{\mathrm{d}}{\mathrm{d}t} + b_2 \frac{\mathrm{d}^2}{\mathrm{d}t^2} + \cdots + b_{n-1} \frac{\mathrm{d}^{n-1}}{\mathrm{d}t^{n-1}} + b_n \frac{\mathrm{d}^n}{\mathrm{d}t^n} \right\} z(t) = u(t) \qquad (6.6.8)$$

The next step is to convert this nth-order differential equation into the n first-order differential equations of the state variable representation.

Suppose that the state variables are defined in the following way:

$$x_1 = z$$

$$x_2 = \dot{x}_1 = \frac{\mathrm{d}z}{\mathrm{d}t}$$

$$x_3 = \dot{x}_2 = \frac{\mathrm{d}^2z}{\mathrm{d}t^2} \tag{6.6.9}$$

$$x_n = \dot{x}_{n-1} = \frac{\mathrm{d}^{n-1}z}{\mathrm{d}t^{n-1}}$$

These are the so-called phase variables (each variable is simply the integral of the previous variable) and the matrix **A** generated by this choice of variables is known as the companion matrix. When these variables are substituted into equation (6.6.8) the following result is obtained:

$$b_0 x_1 + b_1 x_2 + b_2 x_3 + \cdots + b_{n-1} x_n + b_n \dot{x}_n = u \tag{6.6.10}$$

or

$$\dot{x}_n = -\frac{b_0}{b_n} x_1 - \frac{b_1}{b_n} x_2 - \frac{b_2}{b_n} x_3 - \cdots - \frac{b_{n-1}}{b_n} x_n + \frac{1}{b_n} u \tag{6.6.11}$$

The state equations can now be written in matrix form:

$$
\begin{bmatrix} \dot{x}_1 \\ \dot{x}_2 \\ \vdots \\ \dot{x}_{n-1} \\ \dot{x}_n \end{bmatrix} =
\begin{bmatrix}
0 & 1 & 0 & \cdots & 0 \\
0 & 0 & 1 & \cdots & 0 \\
\vdots & & & & \\
0 & 0 & 0 & \cdots & 1 \\
-b_0/b_n & -b_1/b_n & -b_2/b_n & \cdots & -b_{n-1}/b_n
\end{bmatrix}
\begin{bmatrix} x_1 \\ x_2 \\ \vdots \\ x_{n-1} \\ x_n \end{bmatrix} +
\begin{bmatrix} 0 \\ 0 \\ \vdots \\ 0 \\ 1/b_n \end{bmatrix} u
$$

$$\tag{6.6.12}$$

An inverse Laplace transformation can also be applied to equation (6.6.3):

$$\left\{ a_0 + a_1 \frac{\mathrm{d}}{\mathrm{d}t} + a_2 \frac{\mathrm{d}^2}{\mathrm{d}t^2} + \cdots + a_{n-1} \frac{\mathrm{d}^{n-1}}{\mathrm{d}t^{n-1}} + a_n \frac{\mathrm{d}^n}{\mathrm{d}t^n} \right\} z(t) = y(t) \tag{6.6.13}$$

Derivatives of $z(t)$ in this equation can be replaced by the state variables:

$$a_0 x_1 + a_1 x_2 + a_2 x_3 + \cdots + a_{n-1} x_n + a_n \dot{x}_n = y \tag{6.6.14}$$

Finally $\dot{x}_n$ can be removed by substitution of equation (6.6.11):

$$y = \left\{ a_0 - \frac{a_n b_0}{b_n} \right\} x_1 + \left\{ a_1 - \frac{a_n b_1}{b_n} \right\} x_2 + \left\{ a_2 - \frac{a_n b_2}{b_n} \right\} x_3 + \cdots$$

$$+ \left\{ a_{n-1} - \frac{a_n b_{n-1}}{b_n} \right\} x_n + \frac{a_n}{b_n} u \tag{6.6.15}$$

Written in matrix form this becomes

$$y = [c_0 \quad c_1 \quad c_2 \quad \ldots \quad c_{n-1}] \begin{bmatrix} x_1 \\ x_2 \\ \vdots \\ x_{n-1} \\ x_n \end{bmatrix} + a_n/b_n u \tag{6.6.16}$$

where $c_i = a_i - a_n b_i / b_n$.

This equation, taken together with equation (6.6.12), provides a complete state variable representation of the system whose transfer function is $H(s)$. In control theory it is known as the controllable canonical form. The procedure *TFtoSV* shown below converts a rational transfer function H into a state variable form q:

```
PROCEDURE TFtoSV(H: rational; VAR q: statevar);
VAR
    i, j: CARDINAL;
BEGIN
    WITH q DO
        order := H.order;
        FOR i := 1 TO order DO
            FOR j := 1 TO order DO A[i, j] := 0.0 END;
            A[i, order] := -H.b[i-1]/H.b[order];
            IF i > 1 THEN A[i, i-1] := 1.0 END;
            B[i] := H.a[i-1]-H.b[i-1]*H.a[order]/H.b[order];
            C[i] := 0.0;
        END;
        C[order] := 1.0/H.b[order];
        D := H.a[order]/H.b[order];
    END
END TFtoSV;
```

Consider again the filter network shown in figure 6.6. This filter has a transfer function given by

$$H(s) = \frac{1}{2 + (4 \times 10^{-3}s) + (4 \times 10^{-6}s^2) + (2 \times 10^{-9}s^3)}$$

The corresponding state variable representation is

$$\begin{bmatrix} \dot{x}_1 \\ \dot{x}_2 \\ \dot{x}_3 \end{bmatrix} = \begin{bmatrix} 0 & 1 & 0 \\ 0 & 0 & 1 \\ -1 \times 10^9 & -2 \times 10^6 & -2 \times 10^3 \end{bmatrix} \begin{bmatrix} x_1 \\ x_2 \\ x_3 \end{bmatrix} + \begin{bmatrix} 0 \\ 0 \\ 5 \times 10^8 \end{bmatrix} v_{in}$$

$$v_{out} = \begin{bmatrix} 1 & 0 & 0 \end{bmatrix} \begin{bmatrix} x_1 \\ x_2 \\ x_3 \end{bmatrix} + 0 v_{in}$$

Although it describes the same network, this set of equations is apparently quite different from that obtained in the previous section using the physical state variables.

An alternative state variable representation that can be derived from the transfer function is shown below:

$$\begin{bmatrix} \dot{x}_1 \\ \dot{x}_2 \\ \vdots \\ \dot{x}_{n-1} \\ \dot{x}_n \end{bmatrix} = \begin{bmatrix} 0 & \cdots & 0 & 0 & -b_0/b_n \\ 1 & \cdots & 0 & 0 & -b_1/b_n \\ \vdots & & & & \vdots \\ 0 & \cdots & 1 & 0 & -b_{n-2}/b_n \\ 0 & \cdots & 0 & 1 & -b_{n-1}/b_n \end{bmatrix} \begin{bmatrix} x_1 \\ x_2 \\ \vdots \\ x_{n-1} \\ x_n \end{bmatrix} + \begin{bmatrix} c_0 \\ c_1 \\ \vdots \\ c_{n-1} \\ c_n \end{bmatrix} u \qquad (6.6.17)$$

where $c_i = a_i - a_n b_i / b_n$.

$$y = \begin{bmatrix} 0 & \cdots & 0 & 0 & 1/b_n \end{bmatrix} \begin{bmatrix} x_1 \\ x_2 \\ \vdots \\ x_{n-1} \\ x_n \end{bmatrix} + a_n/b_n u \qquad (6.6.18)$$

In control theory this is known as the observable canonical form. As far as determining the time-domain response of networks is concerned the observable and controllable canonical forms are equally suitable.

6.7 Approximate Solution of the State Variable Equations

It has been established that the time-domain behaviour of linear electronic networks can be described by their state variable equations:

$$\dot{x}(t) = A\,x(t) + Bu(t)$$
$$y(t) = C^T x(t) + Du(t) \qquad (6.7.1)$$

The first of these is a set of first-order differential equations which must be solved to give the state vector $x(t)$; then the output $y(t)$ can be determined from the second equation.

An approximate solution of the state equations can be obtained by numerical integration using the rectangular approximation:

$$x(t + T) = x(t) + T\dot{x}(t) \qquad (6.7.2)$$

This approximation is valid provided that the state variables x_i do not change substantially during the time interval T. Combining this equation with the state equations gives

$$\mathbf{x}(t + T) = (\mathbf{I} + T\mathbf{A})\mathbf{x}(t) + T\mathbf{B}u(t) \qquad (6.7.3)$$

In other words, if the state variables and the input are known at a time t, then the values of the variables at a later time $t + T$ can be calculated. Defining two new matrices $\mathbf{P}$ and $\mathbf{R}$ by

$$\begin{aligned} \mathbf{P} &= (\mathbf{I} + T\mathbf{A}) \\ \mathbf{R} &= T\mathbf{B} \end{aligned} \qquad (6.7.4)$$

allows equation (6.7.3) to be written in the form:

$$\mathbf{x}(t + T) = \mathbf{P}\,\mathbf{x}(t) + \mathbf{R}u(t) \qquad (6.7.5)$$

The matrix $\mathbf{P}$ is known as the transition matrix. A procedure *WriteStepResponse* which calculates the step response of a system, given $\mathbf{P}$, $\mathbf{R}$, the coefficients $\mathbf{C}$, D of the original state variable equations and the time step T, is given below:

```
PROCEDURE WriteStepResponse(P: matrix; R, C: vector; D: REAL;
                            n: CARDINAL; T, Tmax: REAL);
VAR
    i, j: CARDINAL;
    t, u, y: REAL;
    x, xx: vector;
BEGIN
    FOR i := 1 TO n DO x[i] := 0.0 END;
    t := 0.0; u := 1.0;
    REPEAT
        y := D*u;
        FOR i := 1 TO n DO
            y := y+C[i]*x[i];
            xx[i] := R[i]*u;
            FOR j := 1 TO n DO xx[i] := xx[i]+P[i, j]*x[j] END
        END;
        printresponse(t, y);
        x := xx;
        t := t+T
    UNTIL t > Tmax;
END WriteStepResponse;
```

A value for T must be chosen before the matrices $\mathbf{P}$ and $\mathbf{R}$ can be set up. Too small a value leads to an unnecessary number of iterations, and for reasons which will be discussed later may result in a loss of accuracy. On the other hand if T is too large the rectangular approximation used to derive equation (6.7.2) is not valid. To satisfy the rectangular approximation, T must be a small fraction, say one hundredth, of the shortest time constant of the network. Now the natural frequencies of the network are given by the poles p_i of the transfer function, so that

$$T \leqslant \frac{1}{100} \min_i \frac{1}{|p_i|} \qquad (6.7.6)$$

If the transfer function is available the poles can be found by factorizing the denominator polynomial. Alternatively the method described in section 6.9 can be used to obtain the characteristic equation, and therefore the poles, from the matrix $\mathbf{A}$.

The procedure *transition* shown below generates the matrices $\mathbf{P}$, $\mathbf{R}$, given $\mathbf{A}$, $\mathbf{B}$ and the time step T:

```
PROCEDURE transition(A: matrix; B: vector; n: CARDINAL;
                     VAR P: matrix; VAR R: vector; T: REAL);
VAR
    i, j: CARDINAL;
BEGIN
    FOR i := 1 TO n DO
        FOR j := 1 TO n DO P[i, j] := T*A[i, j] END;
        P[i, i] := 1.0+P[i, i];
        R[i] := T*B[i]
    END
END transition;
```

Consider the network shown in figure 6.8. The state variable representation of this network, using the physical state variables is

$$\begin{bmatrix} \dot{x}_1 \\ \dot{x}_2 \end{bmatrix} = \begin{bmatrix} -R/L & -1/L \\ 1/C & 0 \end{bmatrix} \begin{bmatrix} x_1 \\ x_2 \end{bmatrix} + \begin{bmatrix} 1/L \\ 0 \end{bmatrix} v_{in}$$

$$v_{out} = [R \quad 0] \begin{bmatrix} x_1 \\ x_2 \end{bmatrix} + 0 v_{in}$$

Substitution of the component values gives the numerical state variable

Figure 6.8
A second-order
bandpass filter

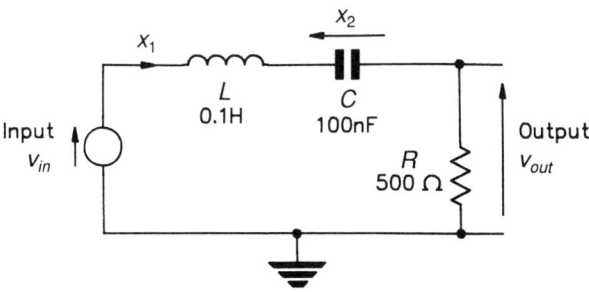

coefficients:

$$\mathbf{A} = \begin{bmatrix} -5000 & -10 \\ 1 \times 10^7 & 0 \end{bmatrix} \qquad \mathbf{B} = \begin{bmatrix} 10 \\ 0 \end{bmatrix}$$

$$\mathbf{C}^T = [500 \quad 0] \qquad D = 0$$

The filter has a resonance frequency of $\omega = 10^4$ rad/s and a Q-factor of 2, so that a suitable choice of the time step would be $T = 10^{-6}$ sec. The **P** and **R** matrices are then given by

$$\mathbf{P} = \begin{bmatrix} 0 \cdot 995 & -10^{-5} \\ 10 & 1 \end{bmatrix} \qquad \mathbf{R} = \begin{bmatrix} 10^{-5} \\ 0 \end{bmatrix}$$

Now, by repeated application of equation (6.7.5) the time-domain response can be determined. If the input v_{in} has a constant value 1 V then a unit-step response will be obtained, as shown in figure 6.9. Two thousand iterations are required to calculate the response from $t = 0$ to $t = 2$ ms using a time step of $1\,\mu\text{s}$, and the computation time on a standard IBM personal computer is approximately 40 sec.

The choice of a suitable value for T is obviously central to this approximate method. Only if T is much less than the shortest time constant of the network is the rectangular approximation valid. It is likely that the response will be required over a time period that is similar to, or greater

Figure 6.9
Unit-step
response of the
bandpass filter

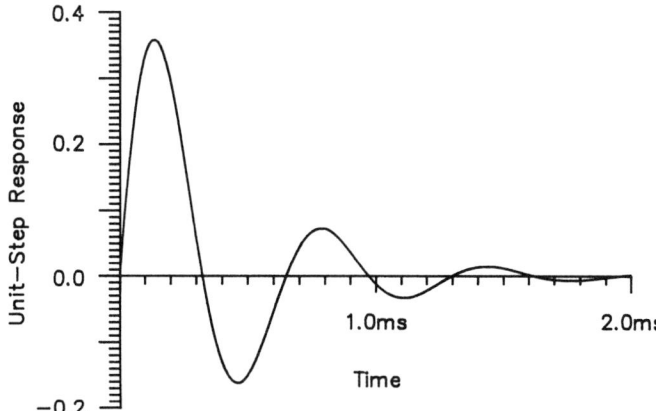

than, the longest time constant of the network. Consequently, if there is a wide spread in the network time constants (or equivalently in the transfer function poles) then a very large number of iterations will be necessary.

Another drawback is the limited accuracy of this method. At first sight it appears that the accuracy can be improved by reducing the size of T, and this is true up to a point. When T becomes very small however, the accuracy starts to get worse again. The reason is to be found in the expression for the diagonal elements of $\mathbf{P}$:

$$P_{ii} = 1 + TA_{ii} \tag{6.7.7}$$

Very small values of T result in a product TA_{ii} that is much less than 1, and although P_{ii} still contains the term TA_{ii} it is at a reduced precision. This effect can clearly be seen in the matrix $\mathbf{P}$ for the bandpass filter of figure 6.8, where $P_{ii} = 1 - 0 \cdot 005 = 0 \cdot 995$. Two decimal digits of accuracy have been lost.

The approximate solution of the state variable equations described here is suitable for determining the response of networks with closely spaced poles to a moderate degree of accuracy. It is simple and is easy to program. If a high degree of accuracy is essential then the matrix exponential solution described in the next section should be employed.

6.8   Matrix Exponential Solution of the State Variable Equations

Before the solution to the vector state equations is derived the simpler case of a scalar state equation will be considered:

$$\dot{x}(t) = ax(t) + bu(t) \tag{6.8.1}$$

This can be rewritten in the form:

$$\frac{\mathrm{d}}{\mathrm{d}t} x(t) - ax(t) = bu(t) \tag{6.8.2}$$

The standard method for solving first-order linear differential equations of this kind involves multiplication of both sides of the equation by an integrating factor. In this case the integrating factor is e^{-at}:

$$e^{-at}\left\{\frac{\mathrm{d}}{\mathrm{d}t} x(t) - ax(t)\right\} = e^{-at}bu(t) \tag{6.8.3}$$

or

$$\frac{\mathrm{d}}{\mathrm{d}t}\left\{x(t)e^{-at}\right\} = e^{-at}bu(t) \tag{6.8.4}$$

Integrating this equation between t and $t + T$ gives

$$x(t + T)e^{-a(t+T)} - x(t)e^{-at} = \int_{t}^{t+T} e^{-a\tau}bu(\tau) \, \mathrm{d}\tau \tag{6.8.5}$$

or

$$x(t + T) = e^{aT}x(t) + \int_t^{t+T} e^{a(t+T-\tau)}bu(\tau)\,d\tau \tag{6.8.6}$$

Provided that the input $u(t)$ remains constant during the interval t to $t + T$, this equation simplifies to

$$x(t + T) = e^{aT}x(t) + \{e^{aT} - 1\}bu(t)/a \tag{6.8.7}$$

This can be written in the form:

$$x(t + T) = px(t) + ru(t) \tag{6.8.8}$$

where $\quad p = e^{aT} \tag{6.8.9}$

$$r = \{e^{aT} - 1\}b/a$$

Equation (6.8.8) shows that if the state variable and the input are known at a time t, then the state variable can be evaluated at a later time $t + T$. No approximations were used to obtain equation (6.8.8), but the input was assumed to be constant over the interval t to $t + T$. The coefficients p and r can both be derived from q, where

$$q = 1 + \frac{(aT)}{2!} + \frac{(aT)^2}{3!} + \cdots$$

$$= \sum_{i=0}^{\infty} \frac{(aT)^i}{(i+1)!} \tag{6.8.10}$$

By writing e^{aT} in the form of a power series:

$$e^{aT} = 1 + \frac{(aT)}{1!} + \frac{(aT)^2}{2!} + \cdots$$

$$= \sum_{i=0}^{\infty} \frac{(aT)^i}{i!} \tag{6.8.11}$$

it can easily be shown that

$$p = 1 + aTq$$
$$r = Tqb \tag{6.8.12}$$

The question that now arises is whether a similar solution can be found for the vector state equations:

$$\dot{\mathbf{x}}(t) = \mathbf{A}\,\mathbf{x}(t) + \mathbf{B}\,u(t) \tag{6.8.13}$$

In the case of the scalar equation an integrating factor of e^{-at} was used. The equivalent integrating factor for the vector state equations is $e^{-\mathbf{A}t}$, but what does this mean? Raising a number to a matrix power seems to make little sense. However, the power series expansion for an exponential given in equation (6.8.11) can be generalized to matrix exponentials:

$$e^{\mathbf{A}t} = \mathbf{I} + \frac{(\mathbf{A}t)}{1!} + \frac{(\mathbf{A}t)^2}{2!} + \cdots$$

$$= \sum_{i=0}^{\infty} \frac{(\mathbf{A}t)^i}{i!}$$

(6.8.14)

It can easily be verified that, interpreted in this way, matrix exponentials have similar properties to normal exponentials. Following the procedure that was used to solve the scalar equation leads to the result:

$$\mathbf{x}(t + T) = \mathbf{P}\,\mathbf{x}(t) + \mathbf{R}\,u(t)$$

(6.8.15)

where

$$\mathbf{P} = e^{\mathbf{A}T} = \mathbf{I} + \mathbf{A}T\mathbf{Q}$$

$$\mathbf{R} = \mathbf{A}^{-1}\{e^{\mathbf{A}T} - \mathbf{I}\}\mathbf{B} = T\mathbf{Q}\mathbf{B}$$

(6.8.16)

and

$$\mathbf{Q} = \sum_{i=0}^{\infty} \frac{(\mathbf{A}T)^i}{(i+1)!}$$

(6.8.17)

This result, like the solution to the scalar equation, is exact but is derived on the assumption that the input $u(t)$ remains constant during the interval t to $t + T$.

The key to solving the vector state equations lies in the evaluation of the matrix $\mathbf{Q}$, whose series expansion is given in equation (6.8.17). In practice, of course, the series must be truncated:

$$\mathbf{Q} = \mathbf{I} + \frac{(\mathbf{A}T)}{2!} + \frac{(\mathbf{A}T)^2}{3!} + \cdots + \frac{(\mathbf{A}T)^m}{(m+1)!}$$

(6.8.18)

What value of m should be used to achieve a particular degree of accuracy? Consider the equivalent scalar power series expansion:

$$q = 1 + \frac{(aT)}{2!} + \frac{(aT)^2}{3!} + \cdots + \frac{(aT)^m}{(m+1)!}$$

(6.8.19)

In this case m should be chosen so that the magnitude of the first missing term is less than the required precision ε:

$$\left| \frac{(aT)^{m+1}}{(m+2)!} \right| < \varepsilon$$

(6.8.20)

If $|aT| \leqslant 1$ this reduces to

$$(m + 2)! > 1/\varepsilon$$

(6.8.21)

To obtain 15 decimal places of accuracy, therefore, $\varepsilon = 10^{-15}$ and a value of $m = 16$ should be used.

Now let the norm of an $n \times n$ matrix $\mathbf{Z}$ be denoted by $\|\mathbf{Z}\|$ where

$$\|\mathbf{Z}\| = \max_{j=1}^{n} \sum_{i=1}^{n} |Z_{ij}|$$

(6.8.22)

It can be proved that the following inequality holds:

$$\| \mathbf{XY} \| \leqslant \| \mathbf{X} \| \, \| \mathbf{Y} \| \qquad (6.8.23)$$

Since $\| \mathbf{I} \| = 1$, it is reasonable to assume that a precision of ε will be obtained when evaluating $\mathbf{Q}$ provided that

$$\left\| \frac{(\mathbf{A}T)^{m+1}}{(m+2)!} \right\| < \varepsilon \qquad (6.8.24)$$

If $\| \mathbf{A}T \| < 1$ then from inequality (6.8.23) it follows that

$$\| (\mathbf{A}T)^{m+1} \| < 1$$

and

$$(m+2)! > 1/\varepsilon \qquad (6.8.25)$$

This result is identical to that obtained for the scalar series expansion, and it shows that matrix powers up to $(\mathbf{A}T)^{16}$ must be included in the expansion to obtain 15 decimal places of accuracy. In practice $\mathbf{Q}$ is not evaluated directly as the power series given in equation (6.8.18). Instead equation (6.8.18) is rearranged to give

$$\mathbf{Q} = \mathbf{I} + \frac{\mathbf{A}T}{2} \left\{ \mathbf{I} + \frac{\mathbf{A}T}{3} \left\{ \cdots \left\{ \mathbf{I} + \frac{\mathbf{A}T}{m} \left\{ \mathbf{I} + \frac{\mathbf{A}T}{m+1} \right\} \right\} \cdots \right\} \right\} \qquad (6.8.26)$$

Although evaluation of $\mathbf{Q}$ in this way involves the same number of matrix multiplications as the direct power series evaluation, it is superior numerically.

One problem still remains: equation (6.8.25) was obtained on the assumption that $\| \mathbf{A}T \| < 1$, but this is by no means always the case. Fortunately any combination of $\mathbf{A}$ and T can be converted to a form suitable for series expansion. Consider the properties of the transition matrix $\mathbf{P}$ evaluated for a time period $2T$:

$$\mathbf{P}(2T) = e^{2\mathbf{A}T} = \{e^{\mathbf{A}T}\}^2 = \mathbf{P}^2(T) \qquad (6.8.27)$$

Now $\mathbf{P}$ is related to $\mathbf{Q}$ by

$$\mathbf{P}(T) = \mathbf{I} + \mathbf{A}T\mathbf{Q}(T) \qquad (6.8.28)$$

so that

$$\begin{aligned} \mathbf{I} + 2\mathbf{A}T\mathbf{Q}(2T) &= \{\mathbf{I} + \mathbf{A}T\mathbf{Q}(T)\}^2 \\ &= \mathbf{I} + 2\mathbf{A}T\mathbf{Q}(T) + T^2\{\mathbf{A}\mathbf{Q}(T)\}^2 \end{aligned} \qquad (6.8.29)$$

This simplifies to

$$2\mathbf{A}T\mathbf{Q}(2T) = 2\mathbf{A}T\mathbf{Q}(T)\{\mathbf{I} + T\mathbf{A}\mathbf{Q}(T)/2\}$$

or

$$\mathbf{Q}(2T) = \mathbf{Q}(T)\{\mathbf{I} + T\mathbf{A}\mathbf{Q}(T)/2\} \qquad (6.8.30)$$

Equation (6.8.30) implies that if $\mathbf{Q}$ has been determined for a time period T, then it can also be determined for a period $2T$ at the expense of two matrix multiplications. It follows that $\mathbf{Q}(2^k T)$ can be obtained from $\mathbf{Q}(T)$, where k is any integer, and that $2k$ matrix multiplications will be necessary.

Evaluation of the matrices $\mathbf{P}$, $\mathbf{R}$ proceeds in three separate stages. First the norm of the matrix $\mathbf{A}$ is determined, and the original value of T is repeatedly halved until $T\| \mathbf{A} \| < 1$. Then $\mathbf{Q}$ is evaluated for this new value of T by a truncated power series (equation 6.8.26). Finally equation (6.8.30) is used to calculate $\mathbf{Q}$ for the original value of T, and the matrices $\mathbf{P}$, $\mathbf{R}$ obtained from equations (6.8.16).

A procedure *transition* for evaluating $\mathbf{P}$, $\mathbf{R}$, given $\mathbf{A}$, $\mathbf{B}$ and T is shown below:

```
PROCEDURE transition(A: matrix; B: vector; n: CARDINAL;
                     VAR P: matrix; VAR R: vector; T: REAL);
VAR
    i, j, k: CARDINAL;
    v, z, t: REAL;
    Q, F: matrix;
BEGIN
    v := 0.0;
    FOR j := 1 TO n DO
        z := 0.0;
        FOR i := 1 TO n DO z := z+ABS(A[i, j]) END;
        IF v < z THEN v := z END;
    END;
    k := 0; t := T;
    WHILE v*t > 1.0 DO INC(k); t := t/2.0 END;
    I(Q, n);
    FOR i := 16 TO 1 BY -1 DO mxa(Q, A, Q, n, t/FLOAT(i+1)) END;
    WHILE k > 0 DO
        mxa(F, A, Q, n, t/2.0);
        mxprod(Q, Q, F, n, n, n);
        DEC(k); t := 2.0*t
    END;
    mxa(P, A, Q, n, T);
    FOR i := 1 TO n DO
        z := 0.0;
        FOR j := 1 TO n DO z := z+Q[i, j]*B[j] END;
        R[i] := T*z
    END
END transition;
```

Three additional procedures are employed in this routine. The procedure *I* sets up an identity matrix, *mxprod* multiplies together two matrices, and *mxa*, which is given below, multiplies together two square matrices, scales the result and adds it to the identity matrix.

```
PROCEDURE mxa(VAR A: matrix; B, C: matrix; n: CARDINAL; t: REAL);
VAR
    i, j, k: CARDINAL;
    z: REAL;
BEGIN
    FOR i := 1 TO n DO
        FOR j := 1 TO n DO
            z := 0.0;
            FOR k := 1 TO n DO z := z+B[i, k]*C[k, j] END;
            A[i, j] := t*z
        END;
        A[i, i] := 1.0+A[i, i]
    END
END mxa;
```

Once the matrices **P**, **R** have been evaluated, the response at times $T, 2T, 3T, \ldots$ can be calculated by repeated application of equation (6.8.15). Step response calculations are particularly straightforward, since $u(t) = 1$, and a suitable procedure *WriteStepResponse* has already been described in the previous section.

Equation (6.8.15) was derived on the assumption that the input remains constant during the interval t to $t + T$. In other words the input function must consist of a sequence of steps. Many input functions are of this type. For example the bipolar pulse shown in figure 6.10 can be represented by

$$
\begin{aligned}
u(t) = & \quad 1 & & \text{for } 0 < t \leqslant T \\
= & -1 & & \text{for } T < t \leqslant 2T \\
= & \quad 0 & & \text{for } t > 2T
\end{aligned}
$$

Continuous input functions must be represented by step approximations, and the calculated responses are no longer exact. Clearly, the accuracy of a step approximation depends on the size of the steps; the smaller the value of T, the higher the accuracy. Consider, for example, the Gaussian pulse shown in figure 6.11. This could be represented to a sufficient accuracy for many purposes by a 40-section step approximation.

Calculation of the transition matrix involves a minimum of 16 matrix multiplications, and represents the major computational effort required to

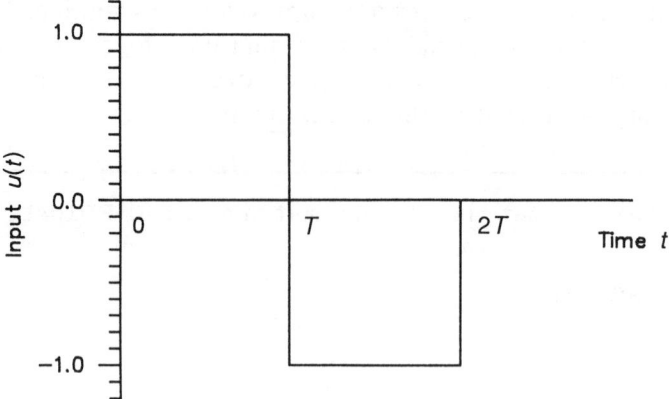

Figure 6.10
A bipolar pulse

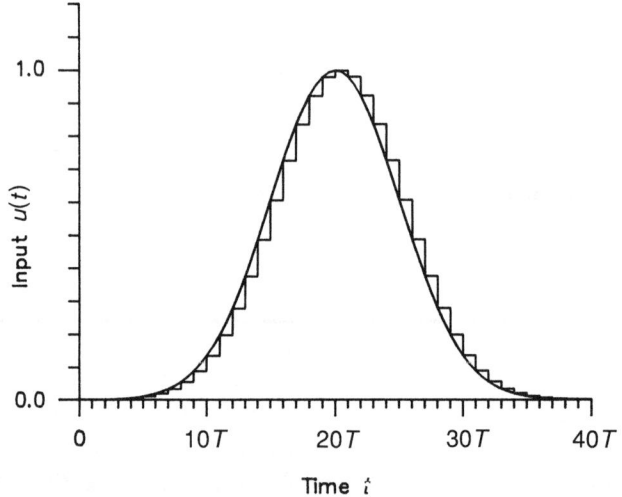

Figure 6.11
Step
approximation
to a Gaussian
pulse

obtain a matrix exponential solution of the state variable equations. The size of the matrices involved is $n \times n$, where n is the order of the network under consideration. Since multiplying together two matrices requires n^3 operations, the time taken to calculate the response of a network is proportional to n^3. On a standard IBM personal computer it takes approximately 15 sec to evaluate an 8×8 transition matrix.

6.9 Transfer Function from the State Variable Representation

It is sometimes necessary to convert the state variable representation of a system to its transfer function, although this is less common than the reverse operation. The following property of Laplace transforms will be used:

$$\mathscr{L} \frac{\mathrm{d}}{\mathrm{d}t} f(t) = sF(s) \tag{6.9.1}$$

(This property was also used in section 6.6 and is valid given zero initial conditions.) With the aid of this result the state variable equations can be transformed to give

$$sX(s) = A\, X(s) + BU(s)$$
$$Y(s) = C^T X(s) + DU(s)$$

(6.9.2)

Combining these equations, and eliminating $X(s)$, gives the transfer function $H(s)$:

$$H(s) = \frac{Y(s)}{U(s)} = C^T(sI - A)^{-1}B + D$$

(6.9.3)

Inversion of the matrix $(sI - A)$ is obviously the key to obtaining $H(s)$ and a procedure for performing this inversion, known as the Souriau–Frame algorithm, will now be described.

The inverse of a matrix is equal to the ratio of its adjoint to its determinant so that

$$(sI - A)^{-1} = \frac{\mathrm{adj}(sI - A)}{|sI - A|}$$

(6.9.4)

If A is of size $n \times n$ then the determinant $|sI - A|$ is an nth order polynomial in s:

$$|sI - A| = b_0 + b_1 s + b_2 s^2 + \cdots + b_{n-1}s^{n-1} + b_n s^n$$

(6.9.5)

This polynomial forms the denominator of the transfer function $H(s)$, and its coefficients can be obtained from the matrix A. Setting the determinant $|sI - A|$ to zero gives the characteristic equation of A:

$$|sI - A| = 0$$

(6.9.6)

The roots of the characteristic equation are the eigenvalues of the matrix A and are also the poles of the transfer function. In section 6.4 it was shown that a change of variable $x = Qz$ transforms a matrix A into a matrix $Q^{-1}A\, Q$. This transformation leads to an identical characteristic equation:

$$\begin{aligned}
|sI - Q^{-1}A\, Q| &= |sQ^{-1}Q - Q^{-1}A\, Q| \\
&= |Q^{-1}(sI - A)\, Q| \\
&= |Q^{-1}||sI - A||Q| \\
&= |sI - A|
\end{aligned}$$

(6.9.7)

The characteristic equation and the eigenvalues are therefore independent of the choice of state variables.

At this point it is necessary to consider the properties of the trace of a matrix. Formally the trace is defined to be the sum of the eigenvalues λ_i of the matrix:

$$\mathrm{tr}\, A = \sum_{i=1}^{n} \lambda_i$$

(6.9.8)

It is easy to show that the trace is also equal to the sum of the elements in the leading diagonal of the matrix:

$$\text{tr } \mathbf{A} = \sum_{i=1}^{n} A_{ii} \tag{6.9.9}$$

and that the trace of the kth power of a matrix is equal to the sum of the kth powers of the eigenvalues:

$$\text{tr } \mathbf{A}^k = \sum_{i=1}^{n} \lambda_i^k \tag{6.9.10}$$

Returning now to the characteristic equation, this can be written either as a polynomial:

$$b_0 + b_1 s + b_2 s^2 + \cdots + b_{n-1} s^{n-1} + b_n s^n = 0 \tag{6.9.11}$$

or equivalently in factorized form:

$$(s - \lambda_1)(s - \lambda_2)(s - \lambda_3)\ldots(s - \lambda_n) = 0 \tag{6.9.12}$$

Equating the coefficients of powers of s in these two equations leads to the following results:

$$b_n = 1$$

$$b_{n-1} = -(\lambda_1 + \lambda_2 + \cdots + \lambda_n)$$
$$= -b_n \text{ tr } \mathbf{A}$$

$$b_{n-2} = \lambda_1\lambda_2 + \lambda_1\lambda_3 + \cdots + \lambda_1\lambda_n + \lambda_2\lambda_3 + \cdots + \lambda_{n-1}\lambda_n$$
$$= (\text{sum of all products of 2 eigenvalues})$$
$$= \{(\lambda_1 + \cdots + \lambda_n)(\lambda_1 + \cdots + \lambda_n) - (\lambda_1^2 + \cdots + \lambda_n^2)\}/2$$
$$= -\{b_{n-1} \text{ tr } \mathbf{A} + b_n \text{ tr } \mathbf{A}^2\}/2$$

$$b_{n-r} = (-1)^r(\text{sum of all products of } r \text{ eigenvalues})$$
$$= (-1)^r\{(-1)^{r-1}b_{n-r+1}(\lambda_1 + \cdots + \lambda_n)$$
$$\qquad -(-1)^{r-2}b_{n-r+2}(\lambda_1^2 + \cdots + \lambda_n^2)$$
$$\qquad +(-1)^{r-3}b_{n-r+3}(\lambda_1^3 + \cdots + \lambda_n^3) \tag{6.9.13}$$
$$\qquad -\vdots$$
$$\qquad +(-1)^{r-1}b_n(\lambda_1^r + \cdots + \lambda_n^r)\}/r$$
$$= -\{b_{n-r+1} \text{ tr } \mathbf{A} + b_{n-r+2} \text{ tr } \mathbf{A}^2 + \cdots + b_n \text{ tr } \mathbf{A}^r\}/r$$

This recursive relationship allows the coefficients b_i to be evaluated in turn, starting with b_n.

The numerator of the transfer function is derived from the adjoint of $(s\mathbf{I} - \mathbf{A})$. Premultiplying both sides of equation (6.9.4) by $|s\mathbf{I} - \mathbf{A}|(s\mathbf{I} - \mathbf{A})$ gives

$$|s\mathbf{I} - \mathbf{A}|\mathbf{I} = (s\mathbf{I} - \mathbf{A}) \text{ adj}(s\mathbf{I} - \mathbf{A}) \tag{6.9.14}$$

Elements in the adjoint matrix are $(n-1) \times (n-1)$ determinants having at most one occurrence of s in each row. Each element is therefore a polynomial in s of order $n-1$, and the complete adjoint matrix can be expressed as a matrix polynomial:

$$\text{adj}(s\mathbf{I} - \mathbf{A}) = \mathbf{R}_0 + \mathbf{R}_1 s + \mathbf{R}_2 s^2 + \cdots \cdots \mathbf{R}_{n-1} s^{n-1} \tag{6.9.15}$$

Substituting this equation and equation (6.9.5) into equation (6.9.14) gives

$$
\begin{aligned}
&\{b_0 + b_1 s + b_2 s^2 + \cdots + b_{n-1} s^{n-1} + b_n s^n\}\mathbf{I} \\
&= (s\mathbf{I} - \mathbf{A})\{\mathbf{R}_0 + \mathbf{R}_1 s + \mathbf{R}_2 s^2 + \cdots + \mathbf{R}_{n-1} s^{n-1}\}
\end{aligned} \tag{6.9.16}
$$

or

$$
\begin{aligned}
&b_0\mathbf{I} + b_1 s\mathbf{I} + b_2 s^2 \mathbf{I} + \cdots + b_{n-1} s^{n-1}\mathbf{I} + b_n s^n \mathbf{I} \\
&= \mathbf{R}_0 s + \mathbf{R}_1 s^2 + \mathbf{R}_2 s^3 + \cdots + \mathbf{R}_{n-1} s^n \\
&\quad - \mathbf{A}\mathbf{R}_0 - \mathbf{A}\mathbf{R}_1 s - \mathbf{A}\mathbf{R}_2 s^2 - \cdots - \mathbf{A}\mathbf{R}_{n-1} s^{n-1}
\end{aligned} \tag{6.9.17}
$$

Equating coefficients of powers of s on both sides of this equation leads to the following results:

$$
\begin{aligned}
\mathbf{R}_{n-1} &= b_n \mathbf{I} \\
\mathbf{R}_{n-2} &= b_{n-1}\mathbf{I} + \mathbf{A}\mathbf{R}_{n-1} \\
&\vdots \\
\mathbf{R}_{n-r} &= b_{n-r+1}\mathbf{I} + \mathbf{A}\mathbf{R}_{n-r+1}
\end{aligned} \tag{6.9.18}
$$

The recursion implicit in this equation for $\mathbf{R}_{n-r}$ can be removed by substituting higher coefficients:

$$
\begin{aligned}
\mathbf{R}_{n-r} &= b_{n-r+1}\mathbf{I} + \mathbf{A}\{b_{n-r+2}\mathbf{I} + \mathbf{A}\{\cdots + \mathbf{A}b_n\mathbf{I}\}\cdots\} \\
&= b_{n-r+1}\mathbf{I} + b_{n-r+2}\mathbf{A} + b_{n-r+3}\mathbf{A}^2 + \cdots + b_n\mathbf{A}^{r-1}
\end{aligned} \tag{6.9.19}
$$

Finally, comparing this result with the expression for the coefficient b_{n-r} (equation 6.9.13) shows that

$$b_{n-r} = -\text{tr}(\mathbf{A}\mathbf{R}_{n-r})/r \tag{6.9.20}$$

This result, together with equations (6.9.18) constitute the Souriau–Frame algorithm for calculating the determinant and adjoint polynomials of $(s\mathbf{I} - \mathbf{A})$:

$$
\begin{aligned}
b_n\quad &= 1 \\
\mathbf{R}_{n-1} &= b_n \mathbf{I} \\
b_{n-1} &= -\text{tr}(\mathbf{A}\mathbf{R}_{n-1})/1 \\
\mathbf{R}_{n-2} &= b_{n-1}\mathbf{I} + \mathbf{A}\mathbf{R}_{n-1} \\
b_{n-2} &= -\text{tr}(\mathbf{A}\mathbf{R}_{n-2})/2 \\
&\vdots \\
\mathbf{R}_{n-r} &= b_{n-r+1}\mathbf{I} + \mathbf{A}\mathbf{R}_{n-r+1} \\
b_{n-r} &= -\text{tr}(\mathbf{A}\mathbf{R}_{n-r})/r
\end{aligned} \tag{6.9.21}
$$

Starting with b_n and $\mathbf{R}_{n-1}$, the coefficients can be evaluated in turn. Each stage of the algorithm involves one matrix multiplication. Since it takes n^3 operations to multiply together two $n \times n$ matrices, the computational effort required to execute the complete algorithm is proportional to n^4.

The last step in obtaining the transfer function is the evaluation of the numerator polynomial coefficients. Combining equations (6.9.4) and (6.9.3) gives

$$H(s) = \frac{\mathbf{C}^T \text{adj}(s\mathbf{I} - \mathbf{A})\mathbf{B}}{|s\mathbf{I} - \mathbf{A}|} + D \qquad (6.9.22)$$

The polynomial forms for the adjoint (equation 6.9.15) and the determinant (equation 6.9.5) can now be substituted into this expression:

$$H(s) = \frac{\mathbf{C}^T(\mathbf{R}_0 + \mathbf{R}_1 s + \cdots + \mathbf{R}_{n-1} s^{n-1})\mathbf{B}}{b_0 + b_1 s + \cdots + b_n s^n} + D \qquad (6.9.23)$$

or

$$H(s) = \frac{\mathbf{C}^T\mathbf{R}_0\mathbf{B} + b_0 D + (\mathbf{C}^T\mathbf{R}_1\mathbf{B} + b_1 D)s + \cdots + b_n D s^n}{b_0 + b_1 s + \cdots + b_n s^n} \qquad (6.9.24)$$

The numerator coefficients are therefore given by

$$\begin{aligned} a_i &= \mathbf{C}^T\mathbf{R}_i\mathbf{B} + b_i D \qquad \text{for } i < n \\ a_n &= b_n D \end{aligned} \qquad (6.9.25)$$

A procedure *SVtoTF* which converts a state-variable description q of a system to its transfer function H is shown below:

```
PROCEDURE SVtoTF(q: statevar; VAR H: rational);
VAR
      i, j, r, n: CARDINAL;
      z: REAL;
      R: matrix;
BEGIN
      n := q.order; H.order := n;
      WITH H DO
          a[n] := 0.0; b[n] := 1.0;
          I(R, n);
          FOR r := 1 TO n DO
              z := 0.0;
              FOR i := 1 TO n DO
                  FOR j := 1 TO n DO z := z+q.C[i]*R[i, j]*q.B[j] END
              END;
```

```
                a[n-r] := z;
                mxprod(R, q.A, R, n, n, n);
                z := 0.0;
                FOR i := 1 TO n DO z := z + R[i, i] END;
                z := -z/FLOAT(r);
                b[n-r] := z;
                FOR i := 1 TO n DO R[i, i] := R[i, i] + z END;
        END;
        FOR i := 0 TO n DO a[i] := a[i]+q.D*b[i] END;
    END
END SVtoTF;
```

This procedure makes use of a routine *I* which sets up an identity matrix, and *mxprod* which multiplies together two matrices. The time taken to convert a 10th-order system on a standard IBM personal computer is appoximately 20 sec.

6.10 Summary

Three techniques have been described for determining the time-domain response of linear networks. Inverse Laplace transformation by partial fraction expansion gives the impulse or step response of a network in analytical form. This method is efficient and should be used wherever possible, but it suffers from numerical instability when the network poles are close or coincident.

Numerical convolution can be used to determine the response to any input waveform, provided that the waveform is bounded in time. The impulse response of the network must be available in anlaytical form, so that numerical convolution may be regarded as a way of extending inverse Laplace transformation to deal with waveforms that are not impulse or step functions. It is a computationally demanding technique, involving a numerical integration at each output point.

An alternative approach is to represent the behaviour of the network by state variable equations, which can then be solved. The state variable equations may be derived directly from the network, or indirectly via the transfer function. Two methods for solving the equations have been described. For networks with closely spaced poles an approximate solution can be obtained very simply. When a high degree of accuracy is essential the matrix exponential solution should be used. This gives the exact results for inputs which consist of a sequence of steps; continuous inputs must be represented by a step approximation.

The transfer function of a network can be derived from the state variable form by the use of the Souriau–Frame algorithm.

6.11 Problems

1 True time delays can only be generated by networks containing distributed elements (such as transmission lines). It is, however, possible to devise rational transfer functions which approximate time delays by means of, for example, Padé's method. The (3,3) Padé approximant to a 1 sec time delay is

$$H(s) = \frac{1 - s/2 + s^2/10 - s^3/120}{1 + s/2 + s^2/10 + s^3/120}$$

Use the partial fraction method to determine the unit-step response of this transfer function over the range $t = 0$ sec to $t = 4$ sec.

2 The (4,5) Padé approximant to a 1 sec delay has an impulse response given by

$$h(t) = 273 \cdot 3 \ e^{-6 \cdot 287t}$$
$$- 57 \cdot 72 \ e^{-3 \cdot 656t} \ \sin\{6 \cdot 544t - 2 \cdot 561\}$$
$$- 329 \cdot 4 \ e^{-5 \cdot 700t} \ \sin\{3 \cdot 210t + 1 \cdot 145\}$$

By a process of numerical convolution determine the response of this filter to a unit-step function, and to a Gaussian pulse of width $0 \cdot 5$ sec centred on $t = 2$ sec:

$$u(t) = 0 \qquad \text{for } t < 0$$
$$u(t) = e^{-2(t-2)^2} \qquad \text{for } 0 \leqslant t \leqslant 4$$
$$u(t) = 0 \qquad \text{for } t > 4$$

3 A bandpass filter has a transfer function given by

$$H(s) = \frac{s}{1 + 2s + s^2}$$

Obtain a state variable representation of this filter. By means of the approximate solution method, calculate the unit-step response at a time $t = 1$ sec, using time steps of $0 \cdot 1$ sec, $0 \cdot 01$ sec and $0 \cdot 001$ sec. Compare your results with the exact value $1/e = 0 \cdot 36787944$. Repeat this calculation using the matrix exponential method with a time step of 1 sec.

4 Figure 6.12 shows a second-order bandpass active filter network. Using the capacitor voltages as the physical state variables, derive a state variable representation of this filter. By means of the matrix exponential method, calculate the unit-step response over the range $t = 0$ ms to $t = 20$ ms, using a time step of $0 \cdot 2$ ms.

Figure 6.12
A Rauch filter

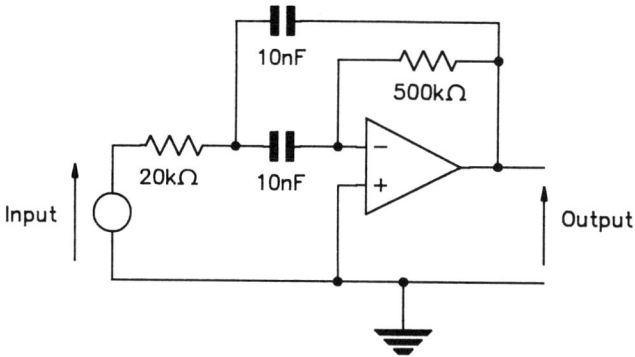

5   The state variable equations given below represent a third-order filter:

$$\begin{bmatrix} \dot{x}_1 \\ \dot{x}_2 \\ \dot{x}_3 \end{bmatrix} = \begin{bmatrix} -1 & 1 & -2 \\ -3 & 2 & -6 \\ -1 & 1 & -3 \end{bmatrix} \begin{bmatrix} x_1 \\ x_2 \\ x_3 \end{bmatrix} + \begin{bmatrix} 1 \\ 1 \\ 0 \end{bmatrix} u$$

$$y = \begin{bmatrix} 0 & 0 & -1 \end{bmatrix} \begin{bmatrix} x_1 \\ x_2 \\ x_3 \end{bmatrix} + 0u$$

Use the Souriau–Frame algorithm to determine the transfer function $H(s) = Y(s)/U(s)$ of this filter.

7 Numerical Integration

7.1 Introduction

A common requirement in electronic network analysis and design is to evaluate the integral of some function over a specified range. In most cases the integration has to be performed numerically, usually because the function can itself only be determined numerically. A typical application of numerical integration is the calculation of the noise-equivalent bandwidth of an electronic filter.

Noise-equivalent bandwidth is a measure of how a filter responds to a white noise input. It is defined to be the bandwidth of the ideal sharp cut-off filter that transmits the same amount of noise as the filter under consideration. Noise-equivalent bandwidth is an important parameter of filters which are to be used in signal recovery applications.

Suppose that the input noise power spectral density is $G(f)$. Then the output spectral density $G'(f)$ of a network with frequency-response function $H(j\omega)$ will be

$$G'(f) = |H(j2\pi f)|^2 G(f) \qquad (7.1.1)$$

Integrating the output noise spectral density over all frequencies gives the total output noise power p:

$$p = \int_0^\infty G'(f)\, \mathrm{d}f$$

$$= \int_0^\infty |H(j2\pi f)|^2 G(f)\, \mathrm{d}f \qquad (7.1.2)$$

White noise has a uniform spectral density so that writing $G(f) = \eta$ gives

$$p = \eta \int_0^\infty |H(j2\pi f)|^2\, \mathrm{d}f$$

$$= \frac{\eta}{2\pi} \int_0^\infty |H(j\omega)|^2\, \mathrm{d}\omega \qquad (7.1.3)$$

Now an ideal sharp cut-off filter has a frequency response which is flat up to ω_0:

$$|H_i(j\omega)| = H_0 \qquad \text{for } 0 \leqslant \omega < \omega_0$$
$$= 0 \qquad \text{for } \omega > \omega_0 \tag{7.1.4}$$

The gains of the ideal and actual filters are assumed to be equal at some frequency; in the case of low-pass filters this frequency is taken to be $\omega = 0$:

$$H_0 = |H(j0)| \tag{7.1.5}$$

Finally, if the filters have equal noise power at their outputs, then $\omega_0 = 2\pi f_0$, where f_0 is the noise-equivalent bandwidth:

$$f_0 = \frac{\displaystyle\int_0^{\infty} |H(j\omega)|^2 \, d\omega}{2\pi |H(j0)|^2} \tag{7.1.6}$$

In a few special cases it is possible to perform the integration analytically. For example, consider the first-order low-pass filter shown in figure 7.1. This filter has a frequency-response function given by

$$H(j\omega) = \frac{1}{1 + j\omega CR} \tag{7.1.7}$$

and the noise-equivalent bandwidth is

$$f_0 = \frac{1}{2\pi} \int_0^{\infty} \frac{1}{1 + \omega^2 C^2 R^2} \, d\omega$$

$$= \frac{1}{2\pi RC} \int_0^{\infty} \frac{1}{1 + x^2} \, dx$$

$$= \frac{1}{2\pi RC} [\arctan x]_0^{\infty}$$

$$f_0 = \frac{1}{4RC} \tag{7.1.8}$$

However, in most cases it will be necessary to perform the integration numerically. Consider the network shown in figure 7.2. This filter has a third-order frequency response function:

$$H(j\omega) = \frac{1 + (1 \cdot 6667 \times 10^{-8} (j\omega)^2)}{1 + (5 \times 10^{-4} j\omega) + (1 \times 10^{-7} (j\omega)^2) + (8 \cdot 3333 \times 10^{-12} (j\omega)^3)}$$

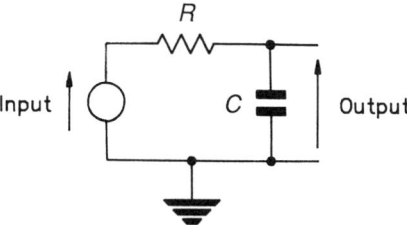

Figure 7.1
A first-order
low-pass filter

Figure 7.2
A third-order
low-pass filter

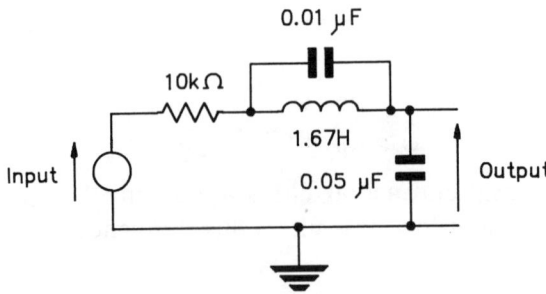

Since the numerator order is one less than the denominator order, the gain at high frequencies will vary inversely with frequency. If the integration is to be performed numerically then a finite upper integration limit ω_i must be chosen. Too low a limit will reduce the accuracy of the result, whereas too high a limit will result in unnecessary computation. A simple error analysis shows that the order of magnitude of the fractional error in the integration result is ω_0/ω_i where ω_0 is the cut-off frequency (the -3 dB point) of the response. In the case of the frequency response function shown above, the cut-off frequency is around 2×10^3 rad/s, so that an upper integration limit of at least 2×10^7 rad/s is necessary to achieve an accuracy of $0 \cdot 01\%$.

7.2 Trapezoidal Integration

Consider the problem of calculating the integral of the function $f(x)$ over the range a to b. Numerical integration methods split the integration range into a number of intervals, estimate the integral of $f(x)$ in each of these intervals, and sum the results. Provided that the intervals are sufficiently small the area under the curve can be approximated by a simple geometrical shape. In trapezoidal integration the range a to b is split into equal approximation intervals, and the area ΔI_i in an interval x_i to x_{i+1} is taken to be the area of the trapezium formed:

$$\Delta I_i = \tfrac{1}{2}(x_{i+1} - x_i)\{f(x_i) + f(x_{i+1})\} \tag{7.2.1}$$

This is illustrated in figure 7.3.

If the range is split into n sections then

$$x_i = a + ih$$

where $h = (b - a)/n$

and the total area under the curve between a and b is given by

$$I = \sum_{i=0}^{n-1} \Delta I_i$$

$$= h\left\{\tfrac{1}{2}f(a) + \sum_{i=1}^{n-1} f(a + ih) + \tfrac{1}{2}f(b)\right\} \tag{7.2.2}$$

Figure 7.3
The trapezium
approximation

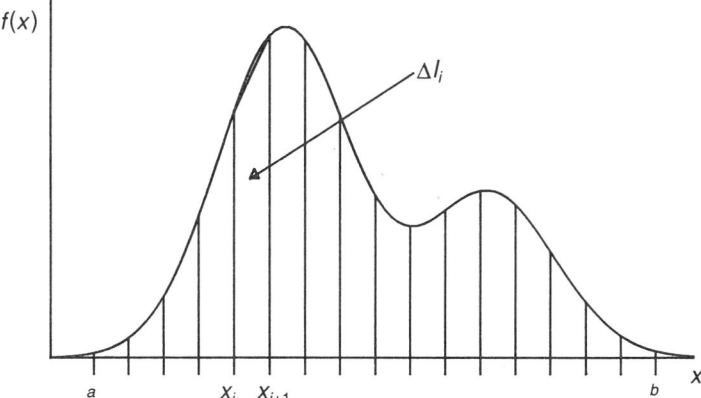

All of the points within the range have equal weights except the end points which have half the weight of the other points.

Before the trapezoidal method can be used to integrate a particular function, a suitable value for the interval h must be chosen. If h is too large then the areas of the individual trapezia will be poor approximations to the area under the curve. On the other hand if h is too small then the amount of computation will become excessive. Also, the accuracy may be reduced as a result of rounding errors in a very large number of additions. In practice h is chosen to be the largest value that gives an acceptable degree of accuracy. An analysis of the error involved in the trapezoidal approximation shows that it is proportional to h^3, and to the second derivative of $f(x)$. However, since the second derivative is not usually accessible, there is no *a priori* way of determining a suitable value for h.

A simple way around this difficulty is to choose an initial value of h which is known to be too large, for example $h = b - a$, and to evaluate the integral using the trapezoidal method. Then h is halved and the integral again evaluated. This process is repeated until two successive values for the integral differ by less than the permissible error. At first sight this method appears to involve twice as many calculations of $f(x)$ as are actually used in the final result. However, by careful programming it is possible to use all evaluations of $f(x)$.

A general integration procedure should accept the function to be integrated as one of its parameters and a suitable type definition for this function would be:

TYPE

```
function = PROCEDURE (REAL): REAL;
```

The procedure *trapezium* shown below integrates the function f over the range *a* to *b*, to an absolute accuracy *eps*:

```
PROCEDURE trapezium(f: function; a, b, eps: REAL): REAL;
VAR
    i, n: CARDINAL;
    dx, s, s0: REAL;
BEGIN
    n := 1;
    s := 0.5*(f(a)+f(b));
    REPEAT
        dx := (b-a)/FLOAT(2*n);
        s0 := s;
        FOR i := 1 TO n DO s := s+f(a+FLOAT(2*i-1)*dx) END;
        n := 2*n
    UNTIL ABS(dx*(2.0*s0-s)) < eps;
    RETURN s*dx
END trapezium;
```

Successively smaller intervals are used until the absolute difference between evaluations is less than the accuracy parameter *eps*.

Trapezoidal integration is straightforward, reliable, and can be satisfactorily applied to many numerical integration problems. When used to calculate the noise-equivalent bandwidth, however, it will almost certainly fail. This is because a satisfactory accuracy can only be obtained if the interval *h* is much less than the cut-off frequency of the filter, while at the same time the upper integration limit must be several orders of magnitude greater than the cut-off frequency. This leads to an excessive number of function evaluations. Trapezoidal integration is not the only method to suffer from this difficulty. Any numerical integration procedure which splits the integration range into sections of equal length is likely to be unsatisfactory for determining noise-equivalent bandwidth.

One way of overcoming these difficulties is to split the frequency range in a logarithmic fashion, and then to apply the integration procedure separately to each of the sub-ranges. For example, in the case of the filter shown in figure 7.2, which has a cut-off frequency of around 2×10^3 rad/s, sub-ranges of $0-10^4$, 10^4-10^5, 10^5-10^6, 10^6-10^7 and 10^7-10^8 rad/s could be used. Under these conditions the procedure *trapezium* gave a result of $500 \cdot 002$ Hz for the noise-equivalent bandwidth. A total of 886 evaluations of the frequency response were required to obtain this result, which is in error by less than $0 \cdot 01\%$.

7.3 Adaptive Integration

Adaptive integration is the name given to integration methods which adapt the approximation interval to the behaviour of the function. Where the function is large in value and varying sharply, the interval is small; where the function is small and varying smoothly the interval is large. Although it is possible to use the trapezoidal approximation adaptively, a more efficient approach uses a polynomial, rather than a straight line, to approximate the function.

Suppose that the function is evaluated at three points x_i, x_{i+1} and x_{i+2}, where x_{i+1} is mid-way between x_i and x_{i+2}. By appropriate choice of coefficients a quadratic polynomial can be fitted to the function at these three points. The area ΔI_i under the quadratic can be derived analytically in terms of its coefficients, and substitution of the relations between the coefficients and the function values yields the following expression for ΔI_i:

$$\Delta I_i = \tfrac{1}{6}(x_{i+2} - x_i)\{f(x_i) + 4f(x_{i+1}) + f(x_{i+2})\} \tag{7.3.1}$$

This result is known as Simpson's rule.

Simpson's rule can be applied non-adaptively to determine an integral by splitting the integration range into an even number of sections. Equation (7.3.1) is then used to determine the area under the curve for pairs of sections. Unfortunately, the additional accuracy that might be expected from the polynomial approximation is not realized in practice. This is because, for a given number of function evaluations, Simpson's rule spans twice the interval of the trapezoidal approximation.

An adaptive integration procedure can be based on the following principle. Suppose that a function is to be integrated over the range a to b. The range is bisected by a new point c, and Simpson's rule used to estimate the integral I_{ab}:

$$I_{ab} = \tfrac{1}{6}(b - a)\{f(a) + 4f(c) + f(b)\} \tag{7.3.2}$$

Also the ranges a to c and c to b are bisected and their integrals I_{ac} and I_{cb} estimated using Simpson's rule. If $I_{ab} = I_{ac} + I_{cb}$ to the required degree of accuracy, then the value $I_{ac} + I_{cb}$ is returned as the value of the integral. Otherwise the procedure is applied recursively to the intervals a to c and c to b, and the sum of these integrals returned.

In the form described above, this integration procedure involves repeated evaluations, at different levels of recursion, of the function at certain points. Obviously this is inefficient. An adaptive integration procedure, incorporating minor modifications to avoid unnecessary function evaluations, is given below:

153

```
PROCEDURE adaptive(f: function; a, b, eps: REAL): REAL;
VAR
    c: REAL;

    PROCEDURE recursive(a, c, b, fa, fc, fb, eps: REAL): REAL;
    VAR
        qab, qac, qcb, d, e, fd, fe: REAL;
    BEGIN
        d := (a+c)/2.0; fd := f(d);
        e := (c+b)/2.0; fe := f(e);
        qab := (b-a)*(fa+4.0*fc+fb)/6.0;
        qac := (c-a)*(fa+4.0*fd+fc)/6.0;
        qcb := (b-c)*(fc+4.0*fe+fb)/6.0;
        IF ABS(qab-qac-qcb) < eps THEN
            RETURN qac+qcb
        ELSE
            RETURN recursive(a, d, c, fa, fd, fc, 0.7*eps) +
                    recursive(c, e, b, fc, fe, fb, 0.7*eps)
        END
    END recursive;

BEGIN
    c := (a+b)/2.0;
    RETURN recursive(a, c, b, f(a), f(c), f(b), eps)
END adaptive;
```

This procedure was used to evaluate the noise-equivalent bandwidth of the filter shown in figure 7.2. With an upper integration limit of 10^8 rad/s, a noise-equivalent bandwidth of $499 \cdot 995$ was obtained, which is accurate to better than $0 \cdot 01\%$. A total of 210 evaluations of the frequency response was required to produce this result.

7.4 Summary

Numerical integration has a number of applications in the analysis of electronic networks. One important application is the calculation of noise-equivalent bandwidths of filters. This involves the integration of the squared modulus of the frequency-response function from zero to a frequency well above the cut-off frequency. Two methods of numerical integration are described.

The trapezoidal method splits the integration range into a large number of equal width intervals, and the area under the curve in each interval is approximated by a trapezium. Summing the individual areas gives the required integral. The accuracy depends on the approximation interval, but there is no way of determining a suitable interval in advance. Instead, the integration is repeated with successively smaller intervals until two results agree to within the required accuracy. Trapezoidal integration is a valuable technique, but is unsuitable for determining noise-equivalent bandwidth because the number of approximation intervals is excessive.

Adaptive integration overcomes this problem by varying the approximation interval according to the behaviour of the function. Simpson's rule can be used as the basis of an adaptive integration procedure. If Simpson's rule gives the integral over the range to a sufficient degree of accuracy, then this value is returned. Otherwise the range is bisected, and the procedure applied recursively to each of the sub-ranges. An algorithm based on this principle is suitable for evaluating noise-equivalent bandwidth.

7.5 Problems

1   A Gaussian distribution $p(x)$, of standard deviation σ, centred on $x = a$, is given by

$$p(x) = \frac{1}{\sigma\sqrt{(2\pi)}} \exp\left(-\frac{(x-a)^2}{2\sigma^2}\right)$$

Determine the proportion of the area of a Gaussian distribution that lies above $a + \sigma$, $a + 2\sigma$, $a + 3\sigma$ and $a + 4\sigma$.

2   Calculate the noise-equivalent bandwidth of a filter whose frequency response function $H(j\omega)$ is given by

$$H(j\omega) = \frac{1 + 0\cdot02381(j\omega)^2}{1 + 0\cdot5j\omega + 0\cdot1071(j\omega)^2 + 0\cdot01190(j\omega)^3 + 5\cdot952 \times 10^{-4}(j\omega)^4}$$

3   The response $g(t)$ of a linear system to a unit-step function input is the integral of its impulse response function $h(t)$:

$$g(t) = \int_0^t h(\tau)\, d\tau$$

Calculate the step response, from $t = 0$ to $t = 2$ sec, in steps of $0\cdot1$ sec, of a fifth-order low-pass Bessel filter whose impulse response is given by

$$h(t) = 20 \cdot 86 \; e^{-3 \cdot 647t}$$
$$+ 6 \cdot 123 \; e^{-2 \cdot 325t} \sin(3 \cdot 571t + 2 \cdot 628)$$
$$+ 27 \cdot 03 \; e^{-3 \cdot 352t} \sin(1 \cdot 743t - 1 \cdot 082)$$

4 Elliptic integrals are important in the design of electronic filters and can only be evaluated numerically. The elliptic integral of the first kind, $u(\phi, k)$, is given by

$$u(\phi, k) = \int_0^{\phi} \frac{d\theta}{\sqrt{(1 - k^2 \sin^2 \theta)}}$$

where $0 \leqslant k \leqslant 1$, and ϕ is in general complex. Calculate the elliptic integrals for $k = 0 \cdot 9$, and $\phi = 0$ to $\phi = 2\pi$ in steps of $\pi/10$.

5 The function $f(x)$ given below is to be integrated between $x = 0$ and $x = 2$.

$$f(x) = x^4 \exp(- x^2)$$

Using the trapezoidal method calculate the integral to accuracies of $10^{-2}, 10^{-3}, \dots 10^{-8}$. In each case determine the number of function evaluations required. Repeat this exercise using adaptive integration.

8 Optimization of Electronic Networks

8.1 Introduction

Optimization is a technique for selecting the component values of an electronic network in such a way as to obtain the best performance. It is essentially a design method and can be used in situations where formal design procedures (such as, for example, filter synthesis) are not available.

Consider, for example, the piecewise-linear waveform shaping network shown in figure 8.1. This network, which consists of a number of parallel limiter sections feeding a virtual ground, converts a 1.0 V triangular waveform into a 1.0 V sinusoid.

A convenient measure of the performance of this network is the total harmonic distortion in the output waveform. Unfortunately, it is necessary to make a number of simplifying assumptions before a formal design procedure can be applied to this network. In particular it is necessary to use an idealized model for the diodes (for example, the current is zero for voltages below $0 \cdot 6$ V; the voltage is $0 \cdot 6$ V for all non-zero currents).

Computer optimization can be used to determine the component values without making any such simplifying assumptions. The output waveform is calculated by analyzing the network at a sequence of input voltages from $-1 \cdot 0$ V to $+1 \cdot 0$ V. At each input voltage the output current of each of the limiter sections is found by numerical solution of the non-linear equations. Component values are then adjusted to minimize the total harmonic distortion in the output waveform.

The output waveform of an optimized five-section shaping network, together with the error between this waveform and a true sinusoid, is shown in figure 8.2. Total harmonic distortion is $0 \cdot 35\%$, which is a substantially lower figure than can be obtained by the use of traditional design methods.

In fact, most network design problems cannot be solved by formal methods. Computers have tended to be used indirectly during the design process, for analyzing the network under consideration and predicting its performance. The reason usually put forward to explain this is that analysis is a science and can readily be automated, whereas design is an art.

Figure 8.1
A piecewise-linear waveform-shaping network

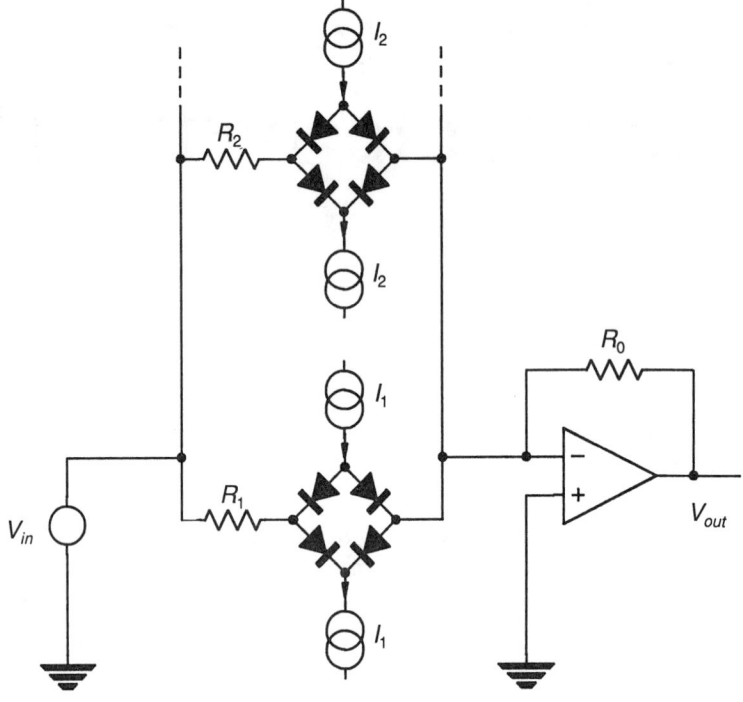

Figure 8.2
Output of the optimized piecewise-linear network

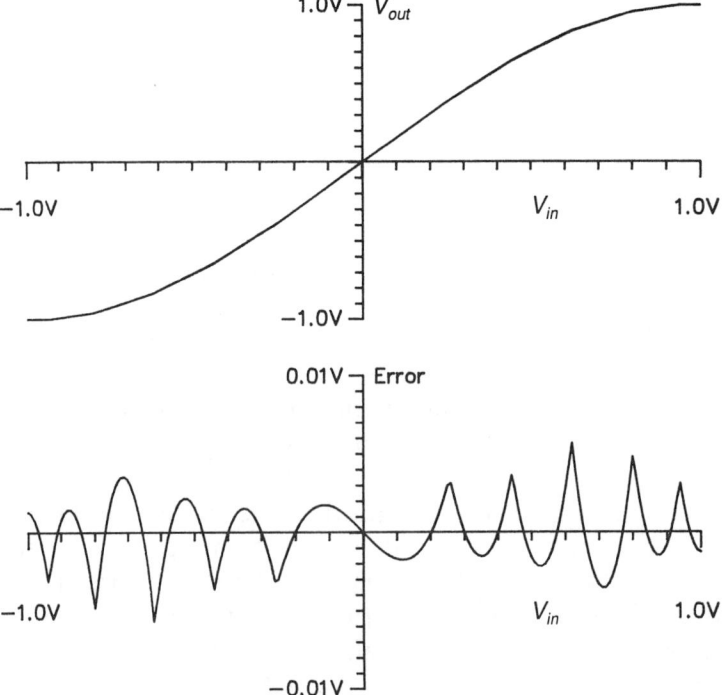

However, the design process consists, at least in part, of considering a large number of possible alternatives, and rejecting all except the best. This is just the type of activity to which computers are well suited.

Computer optimization normally operates on a fixed network configuration and can be considered to be analogous to the manual adjustment of component values in a network prototype. However, manual adjustment of more than about four network values is usually impracticable because of mutual interactions. Computer optimization may succeed where manual adjustment fails because the computer operates much more rapidly, is systematic, and has a perfect memory for the effects of previous component changes.

Any network design that can be analyzed by computer can also be optimized. In effect optimization closes a feedback loop around the network analysis process. Optimization is iterative. At each stage the network is analyzed with a new set of component values and the performance assessed according to some criterion. This performance, together with those measured previously, is used to select component values to be used in the next iteration.

An optimum is considered to have been reached when varying any component value, or any combination of component values, by a small amount, leads to an inferior performance.

Because of the iterative nature of computer optimization it is necessary to make an initial choice of component values. Although the optimization will normally converge, even from a very poor starting point, the number of iterations can be substantially reduced by a good initial choice. Wherever possible, therefore, some approximate design method should be used prior to optimization. In those cases where no approximate design method exists it is possible to make a random selection of the initial component values, but this should be regarded as a last resort.

A block diagram of the complete optimization process is shown in figure 8.3. Although the discussion so far has focussed on the network design problem, optimization has in fact much wider applications in the field of electronic engineering. For example, the coefficients of a transfer function could be chosen to generate a specified frequency response, or the switching frequency chosen to maximize the efficiency of an inverter.

Computer optimization is clearly a powerful tool for electronic network design. It does however have certain drawbacks. Being an iterative process, optimization tends to be computationally expensive. A typical optimization might involve thousands of calculations of the network performance and take several hours on a personal computer or workstation.

A network design will usually have a number of local optima in addition to the single global optimum. In principle the optimization process can converge towards any of the optima and it is impossible to guarantee that the global optimum has been found. However, confidence in a result can be gained by repeating the optimization with different initial values. If the

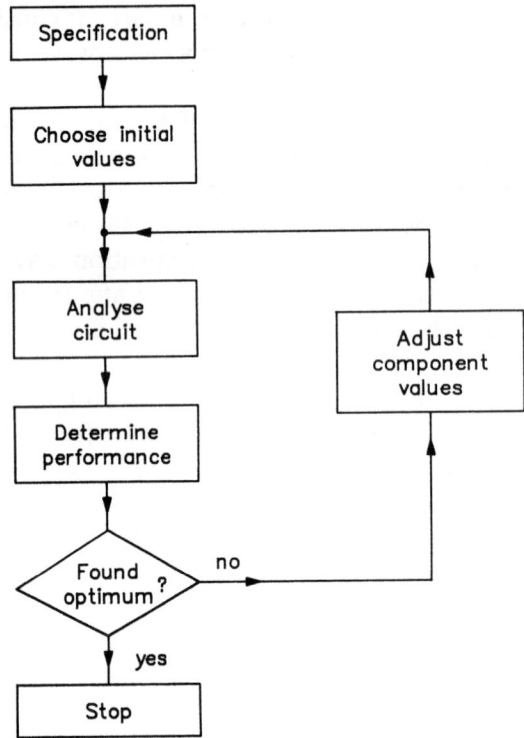

Figure 8.3
The complete optimization process

same optimum is obtained for several starting points then the likelihood is that the global optimum has been found.

Optimization normally operates on a network of fixed topology, with a fixed number of component values being adjusted. This is obviously a severe restriction and leaves the most difficult part of the design, that is choosing the network configuration, to be performed manually. Some attempts have been made to overcome this difficulty by allowing an optimization to make random component additions to an existing network.

Finally, computer optimization gives little insight into the operation of a network. There is a danger that certain design possibilities (such as minor changes in the network configuration) may be overlooked if the technique is applied blindly.

8.2 Function Minimization

Optimization is usually formulated as a minimization problem. It is assumed that the performance of the network under consideration can be embodied in a scalar objective function f which depends on the n variables $x_1, x_2, \ldots x_n$. By convention this objective function is taken to be a measure of how far the actual performance differs from the ideal. Optimization is therefore equivalent to the minimization of f. This in no way restricts the generality of the technique since

$$\text{maximum}(f) = -\text{minimum}(-f) \qquad (8.2.1)$$

As far as possible vector notation will be used and the n variables will be denoted by the column vector $\mathbf{x}$ where

$$\mathbf{x} = \begin{bmatrix} x_1 \\ x_2 \\ \vdots \\ x_n \end{bmatrix} \qquad (8.2.2)$$

Similarly, changes in the variables will be denoted by $\Delta\mathbf{x}$ where

$$\Delta\mathbf{x} = \begin{bmatrix} \Delta x_1 \\ \Delta x_2 \\ \vdots \\ \Delta x_n \end{bmatrix} \qquad (8.2.3)$$

The objective function $f(\mathbf{x})$ can be considered to define a surface in a space of $n + 1$ dimensions. Optimization then involves searching this space for the minimum point on the surface. Unfortunately it is difficult, if not impossible, to visualize surfaces in spaces of high dimensionality. For illustrating the various optimization techniques, functions of two variables which form surfaces in three dimensions will be used.

A computer cannot, of course, observe a complete surface but can only evaluate the objective function at a limited number of discrete points. Consequently all minimization techniques are necessarily iterative. Minimization techniques differ in the way they predict a new point with a lower value of f at each iteration stage.

Minimization methods can be divided into gradient methods, which take account of the local gradient, and search methods which do not. Although it seems obvious that gradient methods should be superior to search methods this is by no means always the case. Determination of the gradient at a point involves $n + 1$ evaluations of the objective function at, and around, this point. The calculated gradient is often of poor accuracy, and may be completely meaningless if the objective function is discontinuous.

When discussing the relative efficiency of various minimization methods it will be assumed that the computational effort is simply proportional to the number of evaluations of the objective function. The computational effort required to execute the minimization algorithm itself will be assumed to be negligible.

8.3 The Objective Function

The objective function measures the difference between the actual performance of a network and the ideal performance. It must be carefully chosen so that it leads to the required specification, while at the same time allowing for efficient minimization. In most cases the objective function is a combination of a number of individual differences or errors, each error corresponding to some part of the specification. For example, consider the problem of fitting the unit-step response $g(t)$ of a network to that of the ideal averaging filter $g_{av}(t)$, where

$$
\begin{aligned}
g_{av}(t) &= 0 && \text{for} && t < 0 \\
&= t && \text{for} && 0 \leqslant t < 1 \\
&= 1 && \text{for} && t \geqslant 1
\end{aligned}
$$

This is illustrated in figure 8.4.

A possible choice for the objective function in this case might be the integral of the difference between the actual and ideal responses:

$$
f = \int_{-\infty}^{\infty} |g(t) - g_{av}(t)| \, dt \tag{8.3.1}
$$

The modulus of the difference is taken to prevent cancellation between positive and negative errors. Unfortunately numerical integration is a computationally expensive operation. In practice a perfectly satisfactory objective function can be generated by measuring the errors e_i between the actual and ideal responses at a large number of points, and combining these errors in some way. Normally the measurement points are evenly spaced throughout the region of interest.

In the case of the averaging filter response there would be no point in calculating errors for $t < 0$ sec since both the actual and ideal responses would be zero. Above $t = 4$ sec the actual response will have settled to a substantially constant value. An appropriate set of error measurements in this case would be evaluated at intervals of $0 \cdot 1$ sec, from 0 sec to 4 sec.

Once the individual errors have been determined, they must be combined into a single objective function. The way that this is done influences both

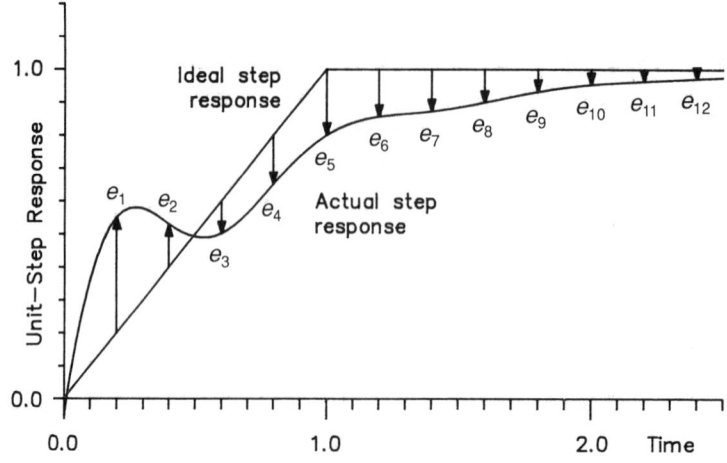

Figure 8.4
Step responses of ideal and actual averaging filters

162

the result of the optimization and the efficiency. Three methods of combining the errors will be considered, namely sum-of-moduli, sum-of-squares and maximum-modulus. Before considering these in detail it is worth looking at the general features of a suitable objective function.

Obviously it must not be possible for a negative error to cancel a positive error. Increasing the modulus of any individual error should lead to a higher objective function. It is also desirable that the objective function should become zero if all the errors are zero; a zero value of the objective function then represents an absolute minimum. Finally the method of combining the errors should absorb a negligible amount of computational effort.

The m individual errors will be represented by a column vector $\mathbf{e}$ where

$$\mathbf{e} = \begin{bmatrix} e_1 \\ e_2 \\ \vdots \\ e_m \end{bmatrix} \tag{8.3.2}$$

8.4 Sum-of-Moduli

One possible method for combining the errors is to take the sum of their moduli:

$$f = \sum_{i=1}^{m} |e_i| \tag{8.4.1}$$

This function satisfies all of the conditions specified above. However, it has the drawback that it does not distinguish between cases where there is one large error and where there are a number of smaller errors. This is illustrated below:

$$f([1001 \cdot 0 \quad 1 \cdot 0 \quad 1 \cdot 0 \quad 1 \cdot 0]) = 1004 \cdot 0$$
$$f([251 \cdot 0 \quad 251 \cdot 0 \quad 251 \cdot 0 \quad 251 \cdot 0]) = 1004 \cdot 0$$

Clearly, in most situations the second set of errors would be preferable to the first.

Another disadvantage of this method for combining the errors is that the derivative of the modulus function is discontinuous around zero. Consequently the derivative of f is discontinuous with respect to the variables $\mathbf{x}$ when any of the individual errors becomes zero. This may lead to problems if gradient methods of minimization are employed.

8.5 Sum-of-Squares

For most purposes the most satisfactory way of combining the errors is by

163

taking the sum of their squares:

$$f = \sum_{i=1}^{m} e_i^2 \qquad (8.5.1)$$

or

$$f = \mathbf{e}^T \mathbf{e} \qquad (8.5.2)$$

The sum-of-squares objective function has the advantage over sum-of-moduli that larger errors are given greater weight and minimization will therefore concentrate on reducing the largest errors first. Provided that the derivatives of the individual errors with respect to the variables $\mathbf{x}$ are continuous, then the sum-of-squares objective function can be used with gradient minimization methods.

A closely related objective function is the sum of even powers of the errors:

$$f = \sum_{i=1}^{m} e_i^p \qquad (8.5.3)$$

This method for combining the errors has the advantage that it places even more emphasis on the largest errors. If a high value of p is used then the objective function effectively becomes dependent only on the largest error and is then very similar in effect to maximum-modulus.

8.6 Maximum-Modulus

In spite of its drawbacks the maximum-modulus objective function is sometimes the appropriate choice:

$$f = \max_{i=1}^{m} |e_i| \qquad (8.6.1)$$

After minimization this will produce an equiripple error function in which there are $n + 1$ equal maxima. Each of these maxima should be smaller than the maximum obtained by using a sum-of-squares objective.

Figure 8.5 illustrates the different results obtained by optimizing with sum-of-squares and with maximum-modulus objective functions. The subject of the optimization is an eighth-order all-pass filter designed to generate a constant $90°$ phase shift between two audio channels over the frequency range 50 Hz to 5 kHz. Such filters are used, for example, in single-sideband modulators. In the upper response a sum-of-squares objective function has been employed and the resulting peak phase error is $0·50°$. When the objective function is changed to maximum-modulus the lower response is obtained which has a peak phase error of $0·33°$.

Unfortunately the maximum-modulus objective function does not 'see' any of the errors except the largest and this can lead to poor efficiency in

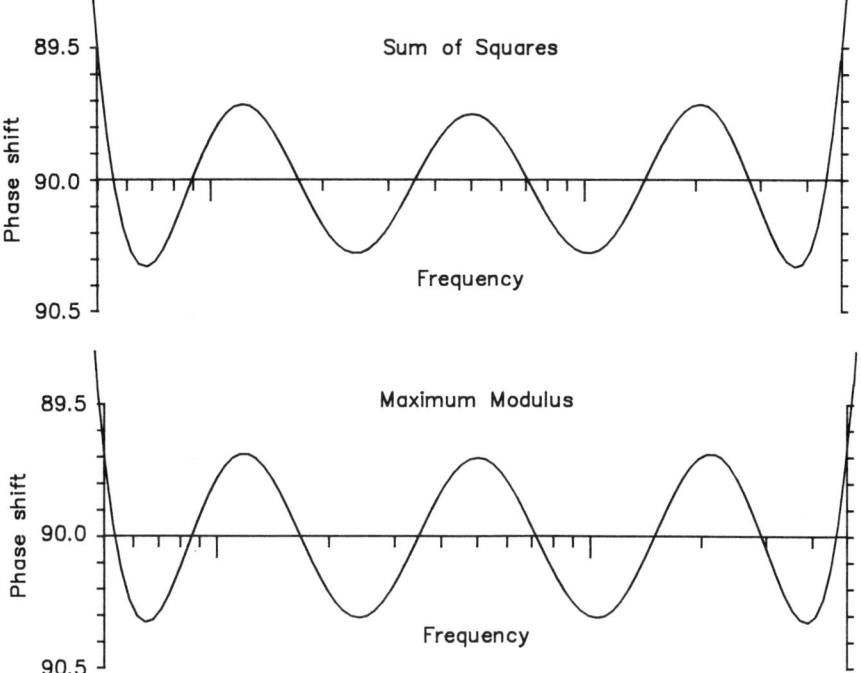

Figure 8.5
Phase responses of optimized phase-shift filters

the minimization process. Also the derivative of the objective function with respect to the variables is discontinuous where the maximum transfers from one error to another. Maximum modulus is therefore an unsuitable objective function for use with gradient minimization methods.

8.7 Error Weighting

Not all parts of a specification are necessarily of equal importance. For example, in the case of the averaging filter considered previously it might be more important to minimize the errors for $t \geqslant 1$ sec than for $t < 1$ sec.

Errors which are of greater importance can be weighted relative to less important errors. In principle each of the errors e_i can be given an individual weight w_i by which it is multiplied before being combined into the objective function. The weighted sum-of-squares objective function is

$$f = \sum_{i=1}^{m} (w_i e_i)^2 \tag{8.7.1}$$

This can be put in vector-matrix form by defining an $m \times m$ diagonal matrix $\mathbf{w}$ which incorporates the squares of the individual weights:

$$\mathbf{w} = \begin{bmatrix} w_1^2 & 0 & \cdots & 0 \\ 0 & w_2^2 & & 0 \\ \vdots & & & \\ 0 & 0 & & w_m^2 \end{bmatrix} \tag{8.7.2}$$

165

Figure 8.6
Effect of
applying
different
weighting
functions

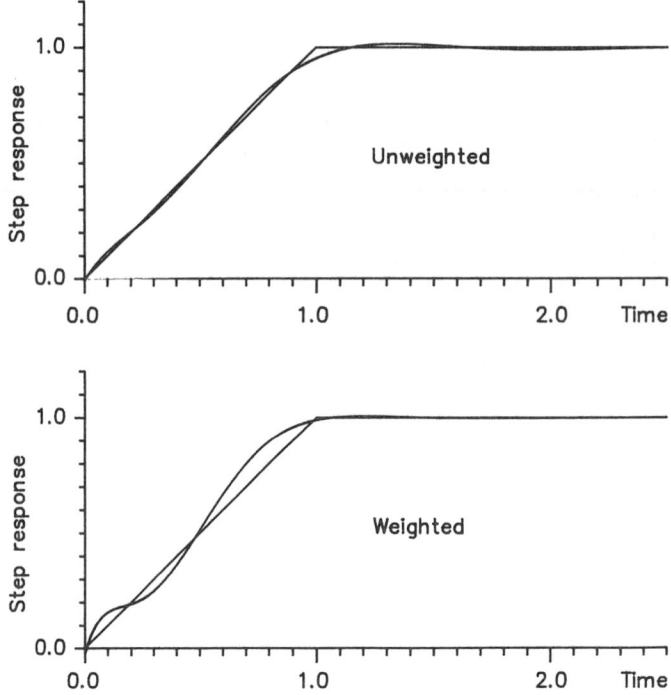

The weighted sum-of-squares objective function then becomes

$$f = \mathbf{e}^T \mathbf{w} \mathbf{e}$$

(8.7.3)

Figure 8.6 illustrates the effect of two different weighting functions applied to the optimization of a third-order approximation to an ideal averaging response. In both cases a sum-of-squares objective function was used, but in the upper response the weighting was uniform, whereas in the lower response the errors for $t \geqslant 1$ sec were weighted by a factor of ten relative to the errors for $t < 1$ sec.

A certain amount of restraint must be used when applying weighting functions. If very large weights are applied to some of the errors then the other errors will be effectively ignored during minimization.

8.8 Optimizing with One Variable

Many problems in electronics involve finding the value of a single component only. Although the multi-variable optimization techniques that will be described later may well work in such cases, special purpose single-variable optimization strategies are more efficient.

Consider, for example, the network of the thermistor thermometer shown in figure 8.7. This network is required to have an output voltage which varies linearly with temperature over the range $0°C$ to $50°C$. It is possible to obtain an approximately linear characteristic because the highly non-

Figure 8.7
A thermistor
thermometer

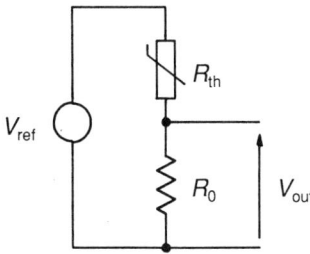

linear behaviour of the thermistor:

$$R_{th} = Ae^{B/T}$$

can to some extent be compensated by the non-linearity of the potential divider ratio with respect to variations in R_{th}:

$$V_{out} = V_{ref}\left(\frac{R_0}{R_0 + R_{th}}\right)$$

The design problem is to determine a value of R_0 that will give the best linearity over the specified temperature range.

Before the design can proceed a suitable measure of the network performance must be devised. The output voltages can be determined for the two temperature extremes and the ideal characteristic taken to be a linear interpolation between these voltages. Errors between the actual and the ideal characteristics, expressed in terms of temperature, can then be evaluated at a large number of intermediate points. In this case it is the maximum deviation of the characteristic from the ideal that is important and an appropriate objective function is therefore the maximum modulus of the individual errors.

Figure 8.8 shows this objective function plotted against the value of R_0. There is an obvious minimum corresponding to a temperature error of

Figure 8.8
Temperature error in the thermistor thermometer

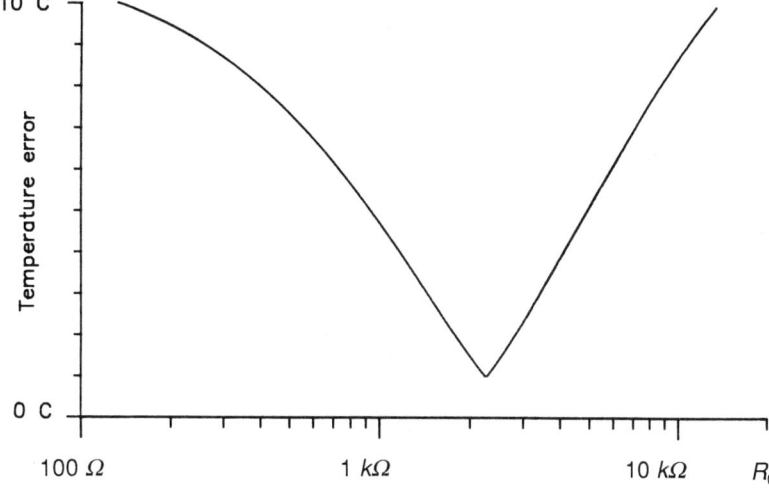

around $1°C$ at a resistor value of about $2 k\Omega$. One way of choosing R_0 would be simply to read the value from the graph. However this gives only an approximate result and is highly inefficient, requiring several hundred evaluations of the objective function.

8.9 Single Variable Approximation Methods

A possible approach to minimizing a function of one variable is to approximate the objective function over a limited range by a polynomial. Cubic polynomials are normally used for this purpose:

$$p(x) = a_0 + a_1 x + a_2 x^2 + a_3 x^3 \qquad (8.9.1)$$

The four coefficients a_0, a_1, a_2, a_3 are chosen to make the polynomial fit the objective function at four points. This involves the solution of four simultaneous linear equations.

The minimum of the cubic polynomial is found by equating the first derivative to zero:

$$x_{min} = \frac{-a_2 \pm \sqrt{(a_2^2 - 3a_1 a_3)}}{3a_3} \qquad (8.9.2)$$

One of the values given by this expression corresponds to the required minimum, the other to an unwanted maximum. The correct value can be selected by evaluating the second derivative:

$$\frac{d^2 p}{dx^2} = 2a_2 + 6a_3 x \qquad (8.9.3)$$

Of course the minimum of the polynomial is only an approximation to the minimum of the objective function, and the process must be repeated, using x_{min} in place of the worst of the original points, until a satisfactory accuracy has been obtained.

Approximation methods are useful in certain applications but cannot be considered to be satisfactory as general single variable optimization methods. The main objection to approximation methods is that they require the objective function to be well-behaved. For a cubic polynomial to be a good approximation, the first, second and third derivatives of the objective function must exist. Clearly approximation methods would be unsuitable for optimizing the thermistor thermometer network discussed previously.

8.10 Single Variable Search Methods

Given an initial interval in which the minimum is known to lie, search methods split the interval into one or more sections and determine which of these sections contains the minimum. The process is repeated, with the

Figure 8.9
Single variable
optimization by
binary search

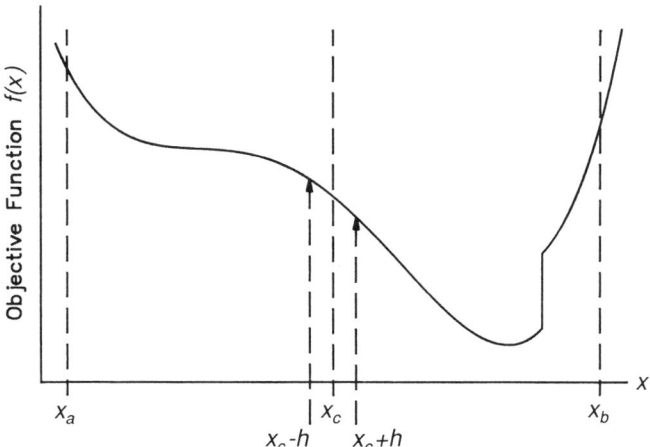

interval becoming progressively smaller, until the minimum has been located to a sufficient degree of accuracy. Since the interval is reduced by a fixed ratio at each iteration, search methods have an entirely predictable convergence. The factor by which the interval is reduced for each evaluation of the objective function is a measure of the efficiency of a search method.

Search methods do not require the objective function to be well-behaved; provided that there is only one minimum within the initial interval then it will be found even if the function is discontinuous. Consider the function shown in figure 8.9.

This function has a single minimum in the range x_a to x_b. If the range is bisected by a point x_c, then it is possible to determine in which half of the range the minimum is to be found by evaluating the objective function at two closely-spaced points on either side of x_c. If $f(x_c - h) < f(x_c + h)$ then the minimum lies in the interval x_a to $(x_c + h)$; otherwise it lies in the interval $(x_c - h)$ to x_b. Provided that h is small compared with $(x_b - x_a)$ then the interval is reduced by a factor of approximately 2 at each iteration. Since two evaluations of the objective function are required the convergence factor per evaluation is $\sqrt{2} = 1 \cdot 414$.

This method, which is obviously closely related to the binary search method for solving non-linear equations, has a satisfactory convergence and will always find a minimum if one exists within the original range. However, the choice of h does present some difficulties. If h is very small then rounding errors in evaluating the objective function may result in the wrong half of the interval being selected, while a larger value will reduce the rate of convergence. These difficulties are overcome in the golden-section search.

8.11 Golden-Section Search

The principle underlying the golden-section search method is that the two evaluations of the objective function are placed so that one of them will also

169

Figure 8.10
The golden-
section method

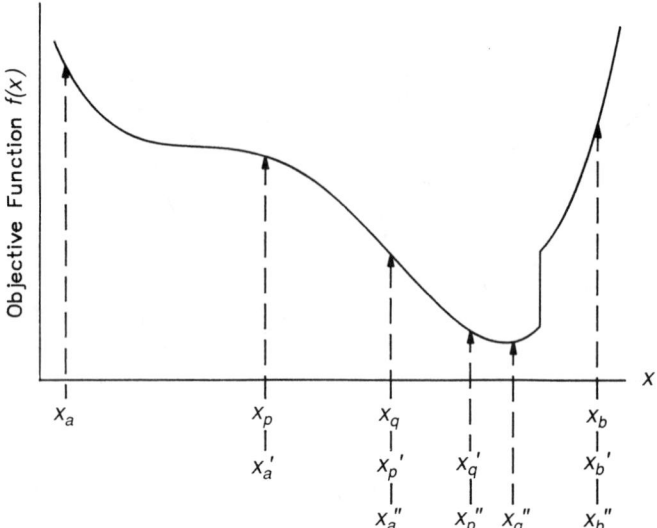

be used in the next iteration. Only one objective function evaluation is therefore necessary at each stage. This is illustrated in figure 8.10.

An initial search interval x_a to x_b is known to contain a minimum. Two points, x_p and x_q, within this interval are chosen for evaluation of the objective function and the relative values of $f(x_p)$ and $f(x_q)$ determine whether the minimum lies in the range x_a to x_q or in the range x_p to x_b.

For reasons of symmetry it will be assumed that the distance between x_a and x_q is equal to that between x_p and x_b. A constant k will be used to denote the reduction in range at each iteration so that

$$x_q - x_a = x_b - x_p = \frac{x_b - x_a}{k} \tag{8.11.1}$$

If $f(x_q) < f(x_p)$ then the minimum lies in the range x_p to x_b. Using x_a' and x_b' to represent the search interval for the next iteration, and x_p' and x_q' to represent the points for evaluation of the objective function, then

$$x_a' = x_p \quad \text{and} \quad x_b' = x_b$$

In order to reduce the number of evaluations the point x_p' is chosen to coincide with x_q where the objective function is already known. Thus only one new object function evaluation, at the point x_q', is required. Applying the same ratio k at this iteration gives

$$x_q' - x_a' = x_b' - x_p' = \frac{x_b' - x_a'}{k} \tag{8.11.2}$$

which can be re-written

$$x_q' - x_p = x_b - x_q = \frac{x_b - x_p}{k} \tag{8.11.3}$$

Combining equations (8.11.1) and (8.11.3), and using the fact that

$$x_b - x_q = (x_b - x_a) - (x_q - x_a) \qquad (8.11.4)$$

gives an expression for the reduction factor k:

$$k^2 - k - 1 = 0 \qquad (8.11.5)$$

The positive root of this equation is

$$k = \frac{1 + \sqrt{5}}{2} = 1 \cdot 618 \qquad (8.11.6)$$

This value, which has the property that $k - 1 = 1/k$, is known as the golden section or golden ratio.

The golden-section method is iterative and some means must be devised for terminating the search when a sufficient degree of accuracy has been achieved. After n iterations the search interval has been reduced by k^{-n}. Consequently the number of iterations can be determined in advance from the initial search interval $x_b - x_a$ and the required accuracy ε:

$$n = \frac{\log(x_b - x_a) - \log(\varepsilon)}{\log(k)} \qquad (8.11.7)$$

In practice it is usually simpler to compare the present search interval with ε at each iteration, and to terminate if it is smaller.

A procedure for performing a golden-section search is given below:

```
PROCEDURE goldensection(f: function; VAR x: REAL; eps: REAL);
CONST
    k = 1.6180339887498948482;
VAR
    xa, xb, xp, xq, fp, fq: REAL;
BEGIN
    xa := 0.0; xb := x;
    xp := xb-(xb-xa)/k; fp := f(xp);
    xq := xa+(xb-xa)/k; fq := f(xq);
    REPEAT
        IF fp > fq THEN
            xa := xp; xp := xq; fp := fq;
            xq := xa+(xb-xa)/k; fq := f(xq)
        ELSE
            xb := xq; xq := xp; fq := fp;
            xp := xb-(xb-xa)/k; fp := f(xp)
    END
```

```
    UNTIL xb-xa < eps;
    x := xp
END goldensection;
```

This procedure was used to optimize the linearity of the thermistor thermometer over the temperature range $0°C$ to $50°C$. The best accuracy was found to occur at a resistance $R_0 = 2·26 \, k\Omega$; with this value of resistance the peak temperature error of the thermometer was $0·97°C$.

8.12 Optimizing in N Dimensions

Methods for optimizing a function of several variables fall into two classes: search methods and gradient methods.

In search methods, no attempt is made to determine the local gradient. Instead, at each iteration a new point is generated which, in the light of previous evaluations of the objective function, might be expected to approach the optimum.

Conceptually the simplest search method consists of adjusting each of the variables in turn, while the other variables remain fixed. A complete iteration involves conducting n one-dimensional searches along the individual variable directions. Unfortunately, interactions between the

Figure 8.11
One-variable-at-a-time optimization

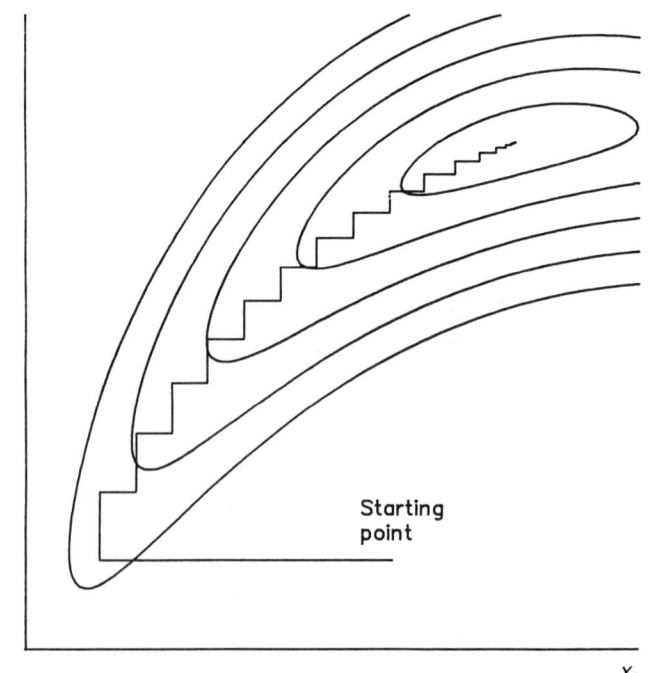

variables usually render this method impracticable. The problem is illustrated for two variables x_1 and x_2 in figure 8.11.

Optimization follows a zig-zag path and none of the individual steps are in the direction of the optimum. The pattern search method of Hooke and Jeeves attempts to improve the efficiency by performing two types of move. The first part of each iteration consists of exploratory moves along the individual variable directions. Taken together, these n exploratory moves define a direction which should point towards the optimum. Pattern moves are then made along this new direction until the lowest value of the objective function has been found.

Although these methods are attractive from the point of view of simplicity, in practice they become increasingly inefficient compared with more sophisticated methods such as simplex, as the number of variables becomes large. The simplex optimization method is described in detail in section 8.13.

Gradient methods make use of the local gradient to predict the direction and distance of the optimum from the present position. A Taylor expansion of the objective function f in terms of the variables $\mathbf{x}$ around the point $\mathbf{x}_0 + \Delta\mathbf{x}$ gives

$$
\begin{aligned}
f(\mathbf{x}_0 + \Delta\mathbf{x}) = f(\mathbf{x}_0) + \sum_{i=1}^{n} \frac{\partial f}{\partial x_i} \Delta x_i \\
+ \frac{1}{2} \sum_{i=1}^{n} \sum_{j=1}^{n} \frac{\partial^2 f}{\partial x_i \, \partial x_j} \Delta x_i \, \Delta x_j \\
+ \cdots
\end{aligned}
\tag{8.12.1}
$$

or

$$
f(\mathbf{x}_0 + \Delta\mathbf{x}) = f(\mathbf{x}_0) + \mathbf{g}^T \Delta\mathbf{x} + \tfrac{1}{2} \Delta\mathbf{x}^T \mathbf{H} \, \Delta\mathbf{x} + \cdots
\tag{8.12.2}
$$

where $\mathbf{g}$ is the Jacobian vector:

$$
\mathbf{g} =
\begin{bmatrix}
\dfrac{\partial f}{\partial x_1} \\[2mm]
\dfrac{\partial f}{\partial x_2} \\[2mm]
\vdots \\[2mm]
\dfrac{\partial f}{\partial x_n}
\end{bmatrix}
\tag{8.12.3}
$$

and $\mathbf{H}$ is the Hessian matrix:

$$\mathbf{H} = \begin{bmatrix} \dfrac{\partial^2 f}{\partial x_1^2} & \dfrac{\partial^2 f}{\partial x_1\,\partial x_2} & \cdots & \dfrac{\partial^2 f}{\partial x_1\,\partial x_n} \\[2ex] \dfrac{\partial^2 f}{\partial x_2\,\partial x_1} & \dfrac{\partial^2 f}{\partial x_2^2} & & \\[2ex] \vdots & & & \\[2ex] \dfrac{\partial^2 f}{\partial x_n\,\partial x_1} & & & \dfrac{\partial^2 f}{\partial x_n^2} \end{bmatrix} \qquad (8.12.4)$$

Only rarely can the Jacobian and Hessian be evaluated analytically and in practice they must be determined by some numerical procedure.

Gradient methods which make use of the Jacobian vector only are known as first-order methods; methods which use both Jacobian and Hessian are termed second-order methods. The steepest-descent optimization technique, which is described in section 8.14, is a first-order gradient method.

When a change is made in one of the variables it will in general affect all of the errors e_i. However, once these errors have been combined into a single objective function all the information concerning their individual variations will have been lost. The least-squares optimization procedure, which is described in section 8.15, operates directly on the individual errors and therefore does not suffer from this loss of information. As a result, least-squares is potentially more efficient than methods which operate on the objective function alone. Unfortunately least-squares cannot be considered to be a general optimization method because it is limited to applications where the objective function is of the sum-of-squares form. Being a gradient method it should only be applied to well-behaved error functions.

8.13 Simplex Optimization

The simplex method is generally accepted to be the most successful non-gradient search method. Simplex is, in fact, not a single method but a family of closely related techniques; the variant that will be discussed here was developed by Nelder and Mead. The name derives from the simplest solid in n-dimensional space. A two-dimensional simplex is a triangle and a three-dimensional simplex is a tetrahedron. In general an n-dimensional simplex has $n + 1$ vertices.

Starting with an initial simplex in the n-dimensional variable space, the objective function is evaluated at each of the $n + 1$ vertices. It is a reasonable assumption that the optimum lies in a direction away from the vertex with the highest objective function and towards the other vertices. If a new point can be found with a lower function value then this replaces the original vertex with the highest value.

At each iteration the simplex changes in shape and size, and gradually adapts itself to the local contours of the objective function. As optimization progresses the simplex tends to become smaller and the search can be terminated once the simplex has contracted to a size that is less than the required accuracy. The initial simplex should be chosen to contain the minimum if at all possible, although the algorithm does not confine the search to the original volume. A detailed description of the Nelder and Mead form of the simplex method will now be given.

The first step is to evaluate the objective function at the $n + 1$ vertices of the simplex. Then the vertices $\mathbf{x}^h$, $\mathbf{x}^s$ and $\mathbf{x}^l$ corresponding to the highest, second highest and lowest values of the objective function can be identified. A new point, the centroid $\mathbf{x}^c$ of the vertices excluding $\mathbf{x}^h$, is constructed:

$$\mathbf{x}^c = \frac{1}{n} \sum_{\substack{i=1 \\ i \neq h}}^{n+1} \mathbf{x}^i \tag{8.13.1}$$

The highest point $\mathbf{x}^h$ is now inverted through $\mathbf{x}^c$ to produce a new point $\mathbf{x}^r$:

$$\mathbf{x}^r = (1 + \alpha)\mathbf{x}^c - \alpha\mathbf{x}^h \tag{8.13.2}$$

where α is a positive constant. Thus $\mathbf{x}^r$ lies on an extension of a line from $\mathbf{x}^h$ to $\mathbf{x}^c$, and if α is taken to be $1 \cdot 0$ then the distance between $\mathbf{x}^h$ and $\mathbf{x}^c$ equals that between $\mathbf{x}^c$ and $\mathbf{x}^r$.

If $f(\mathbf{x}^r) < f(\mathbf{x}^l)$ then the new point is superior to all of the simplex vertices. Since a favourable direction has been found the search is extended to a further point $\mathbf{x}^e$:

$$\mathbf{x}^e = \gamma\mathbf{x}^r + (1 - \gamma)\mathbf{x}^c \tag{8.13.3}$$

where γ is a positive constant which is greater than unity and is usually taken to be 2.0. The highest point is then replaced by whichever of $\mathbf{x}^e$ and $\mathbf{x}^r$ has the lowest objective function, and the process is restarted with this new simplex.

If $f(\mathbf{x}^r) \geqslant f(\mathbf{x}^l)$ but $f(\mathbf{x}^r) < f(\mathbf{x}^s)$ then some improvement has been gained, but not sufficient to warrant extending the search. The highest point $\mathbf{x}^h$ is replaced by $\mathbf{x}^r$ and the process is restarted.

If $f(\mathbf{x}^r) \geqslant f(\mathbf{x}^s)$ but $f(\mathbf{x}^r) < f(\mathbf{x}^h)$ then only a slight improvement has been gained. It may be that a better point has been overshot and so a contraction from the reflected point towards the centroid is performed giving a new point $\mathbf{x}^k$:

$$\mathbf{x}^k = \beta\mathbf{x}^r + (1 - \beta)\mathbf{x}^c \tag{8.13.4}$$

where β is a constant which lies between 0 and 1, and is usually taken to be $0 \cdot 5$. The highest point $\mathbf{x}^h$ is replaced by whichever of $\mathbf{x}^k$ and $\mathbf{x}^r$ has the lowest objective function and the process is restarted.

Finally, if $f(\mathbf{x}^r) \geqslant f(\mathbf{x}^h)$ then no improvement has resulted from moving in what was expected to be a favourable direction. This is probably because

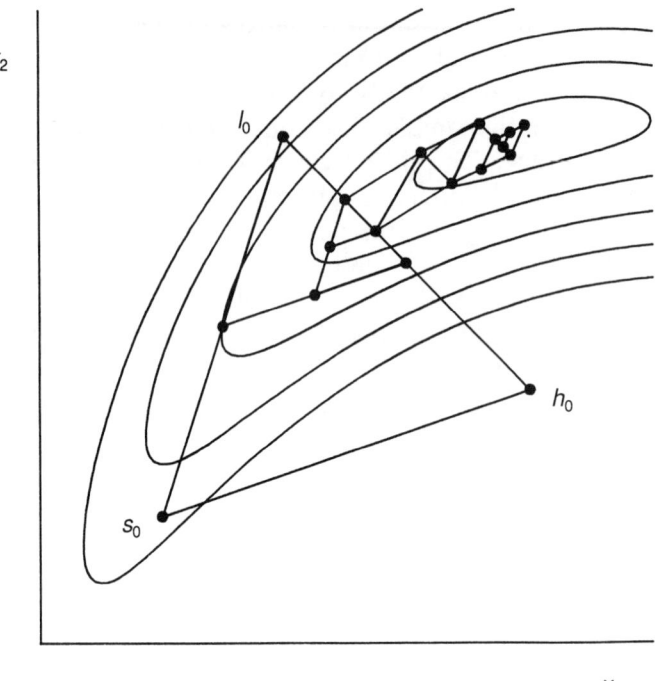

Figure 8.12
Simplex
optimization

the optimum lies within the simplex. A contraction of the whole simplex towards the lowest point is performed by applying the operation:

$$\mathbf{x}^i = \delta \mathbf{x}^i + (1 - \delta)\mathbf{x}^l \qquad (8.13.5)$$

to each of the vertices $\mathbf{x}^i$. δ is a constant which lies between 0 and 1. Small values of δ might be expected to improve the rate of convergence, but if the contraction is too severe then there is a danger that an optimum might be missed. A typical value for δ is $0 \cdot 5$. The optimization process is then restarted with the contracted simplex.

Figure 8.12 illustrates the operation of the simplex method for a function of two variables x_1 and x_2. The function has the form of a curved valley.

General-purpose optimization procedures should accept the objective function as a parameter. A suitable type definition for the objective function is given below:

```
TYPE
    index = [0..maxsize];
    vector = ARRAY index OF REAL;
    objective = PROCEDURE(vector, CARDINAL): REAL;
```

The procedure *simplex* shown below optimizes an objective function f of

176

the variables *x* to an accuracy *eps*:

```
PROCEDURE simplex(f: objective; VAR x: vector; n: CARDINAL;
                  eps: REAL);
CONST
    alpha = 1.0; beta = 0.5; gamma = 2.0; delta = 0.5;
TYPE vertex = RECORD
        coord: vector;
        value: REAL
    END;
VAR
    i, j, lo, se, hi: CARDINAL;
    si: ARRAY [1..maxsize] OF vertex;
    xc, xr, xe, xk: vertex;
    q: REAL;
BEGIN init;
    LOOP lo := 1; hi := 1;
        FOR j := 1 TO n+1 DO
            IF si[j].value > si[hi].value THEN hi := j END;
            IF si[j].value < si[lo].value THEN lo := j END
        END;
        se := lo;
        FOR j := 1 TO n+1 DO
            IF (j <> hi) AND
            (si[j].value > si[se].value) THEN se := j END
        END;
        IF quit() THEN EXIT END;
        FOR j := 1 TO n DO
            q := 0.0;
            FOR i := 1 TO n+1 DO
                IF i <> hi THEN q := q+si[i].coord[j] END
            END;
            xc.coord[j] := q/FLOAT(n)
        END;
        line(xr, 1.0+alpha, xc, si[hi]);
        IF xr.value < si[lo].value THEN
            line(xe, gamma, xr, xc);
            IF xe.value < xr.value THEN si[hi] := xe
            ELSE si[hi] := xr END
```

177

```
            ELSIF xr.value < si[se].value THEN si[hi] := xr;
            ELSIF xr.value < si[hi].value THEN
                line(xk, beta, xr, xc);
                IF xk.value < xr.value THEN si[hi] := xk
                ELSE si[hi] := xr END
            ELSE
                FOR i := 1 TO n+1 DO
                    IF i <> lo THEN
                        line(si[i], delta, si[i], si[lo])
                    END
                END
            END
        END;
        x := si[lo].coord
END simplex;
```

The vertices of the simplex and the points x^c, x^r, x^e, x^k are represented by a data structure *vertex* which contains both the coordinates and the corresponding value of the objective function.

A procedure *line*, which is local to *simplex*, generates a new point x on a straight line joining two existing points *x1* and *x2*, and evaluates the objective function at this new point. It is used to construct the inverted, extended and contracted points, and also to perform the general contraction towards the lowest point.

```
PROCEDURE line(VAR x: vertex; a1: REAL; x1, x2: vertex);
VAR
    i: CARDINAL;
BEGIN
    FOR i:=1 TO n DO
        x.coord[i] := a1*x1.coord[i]+(1.0-a1)*x2.coord[i]
    END;
    x.value := f(x.coord, n)
END line;
```

Before optimization can proceed a simplex must be constructed which contains the starting point (that is, the initial estimate of the position of the optimum). This function is performed by the procedure *init*. Although the choice of initial vertex positions is not particularly critical, it is important that they do not all lie on a single hyperplane, otherwise all operations performed within the simplex algorithm will lead to further points which also lie on this hyperplane, and the optimization will fail. This can readily be understood by considering the case of two variables. The simplex should be a triangle, but if the three vertices lie on a straight line then the search will be confined to points on the line.

Ideally the shape of the initial simplex should reflect the contours of the objective function. Unfortunately it is not possible to achieve this in practice and the dimensions of the simplex along each axis are usually chosen to be proportional to the coordinates of the starting point. This approach cannot be used if any of the coordinates of the starting point are zero. The procedure *init*, which is local to *simplex*, is shown below:

```
PROCEDURE init;
CONST
    e = 0.5;
VAR
    i: CARDINAL;
BEGIN
    FOR i := 1 TO n DO
        si[i].coord := x;
        si[i].coord[i] := (1.0-e)*x[i];
        si[n+1].coord[i] := (1.0+e)*x[i]
    END;
    FOR i:=1 TO n+1 DO si[i].value:=f(si[i].coord, n) END
END init;
```

Each of the first n vertices are generated by moving from the starting point along one of the axes; the $(n + 1)$th vertex is chosen so that the centroid of the simplex vertices coincides with the starting point.

Termination of the optimization process will usually depend on whether the simplex has contracted to a sufficient degree, and is controlled by the local procedure *quit* which returns a Boolean result. A convenient approximate measure of the size of the simplex is the distance between the lowest and highest vertices. In the procedure *quit* shown below, the size of the simplex relative to the coordinate value along each axis is compared with the accuracy parameter *eps*. Optimization will be terminated only when the test succeeds for all n axes. This procedure will fail to terminate the optimization if any of the coordinates tends towards zero, and in such cases an alternative criterion must be used.

```
PROCEDURE quit(): BOOLEAN;
VAR
    q1, q2: REAL;
    j: CARDINAL;
BEGIN
    FOR j := 1 TO n DO
        q1 := si[lo].coord[j];
        q2 := si[hi].coord[j];
        IF ABS((q1-q2)/q1) > eps THEN RETURN FALSE END
    END;
    RETURN TRUE
END quit;
```

8.14 Steepest-Descent Optimization

The steepest-descent method is the simplest of the gradient techniques. It is based on the obvious principle that the gradient of the objective function will indicate the direction of the most rapid decrease in the objective function. A search can therefore be conducted along this direction until a minimum is found, and the process repeated as necessary.

Expanding the objective function as a first-order Taylor series around the point $\mathbf{x}_0 + \Delta\mathbf{x}$ gives

$$f(\mathbf{x}_0 + \Delta\mathbf{x}) = f(\mathbf{x}_0) + \sum_{i=1}^{n} \frac{\partial f}{\partial x_i} \Delta x_i \qquad (8.14.1)$$

or using vector notation:

$$f(\mathbf{x}_0 + \Delta\mathbf{x}) = f(\mathbf{x}_0) + \mathbf{g}^T \cdot \Delta\mathbf{x} \qquad (8.14.2)$$

Thus the change Δf in the objective function resulting from the change in variables $\Delta\mathbf{x}$ is given by

$$\Delta f = \mathbf{g}^T \cdot \Delta\mathbf{x} \qquad (8.14.3)$$

We wish to choose $\Delta\mathbf{x}$ in such a way as to maximize the magnitude of Δf while keeping the sign of Δf negative. Now the scalar product can be written in the form

$$\Delta f = \mathbf{g}^T \cdot \Delta\mathbf{x} = |\mathbf{g}| \, |\Delta\mathbf{x}| \cos \theta \qquad (8.14.4)$$

where θ is the angle between the vectors. It follows that for given magnitudes $|\mathbf{g}|$ and $|\Delta\mathbf{x}|$ the greatest reduction in the objective function occurs when $\theta = \pi$, in other words when $\Delta\mathbf{x}$ is in the direction of the negative gradient $-\mathbf{g}$. Consequently $-\mathbf{g}$ defines the direction of steepest descent.

Since it is only the direction of $-\mathbf{g}$ that is of significance, a unit vector $\mathbf{u}$ in the direction of steepest descent is constructed:

$$\mathbf{u} = -\frac{\mathbf{g}}{|\mathbf{g}|} \qquad (8.14.5)$$

so that

$$\Delta \mathbf{x} = \lambda \mathbf{u} \qquad (8.14.6)$$

where λ is positive, but is otherwise undetermined. A one-dimensional search must be conducted to find the value of λ which minimizes $f(\mathbf{x}_0 + \lambda \mathbf{u})$.

Unfortunately, as can be seen from figure 8.13 the steepest-descent direction does not necessarily point directly towards the optimum. In fact, successive steepest-descent directions are at right angles. This can be seen from the fact that following a one-dimensional search for the best value of λ the gradient of f with respect to λ is zero:

$$\frac{\partial}{\partial \lambda} f(\mathbf{x} + \lambda \mathbf{u}) = \mathbf{g}^T(\mathbf{x} + \lambda \mathbf{u}) \cdot \mathbf{u} = 0 \qquad (8.14.7)$$

Consequently the optimization follows a zig-zag path; the more elliptical the contours, the slower the convergence. Correct scaling of the variables can

Figure 8.13
Steepest-descent
optimization

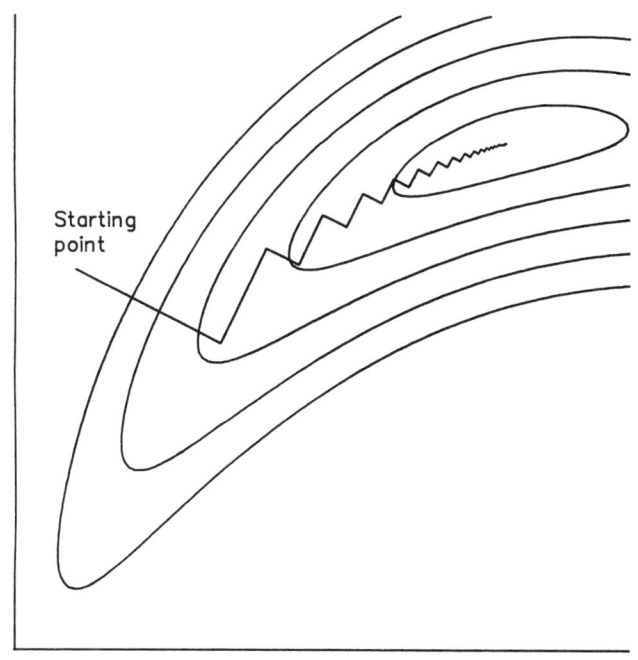

result in a substantial improvement in efficiency of the steepest-descent method.

The n partial derivatives of the objective function must be determined numerically:

$$\frac{\partial f}{\partial x_i} \approx \frac{f(x_1, \dots x_i + \varepsilon_i, \dots x_n) - f(x_1, \dots x_i, \dots x_n)}{\varepsilon_i} \qquad (8.14.8)$$

This involves $(n + 1)$ evaluations of the objective function. Although this appears to be a fairly straightforward procedure, it may be difficult to select suitable values for the increments ε_i. Large values of ε_i lead to the above expression being a poor approximation because of the effect of neglecting the higher derivatives. On the other hand, very small values of ε_i will give a poor accuracy because the calculated derivatives will be the results of subtractions of nearly equal quantities. Normally the magnitude of the ε_i will be related to the corresponding x_i. For example a reasonable choice might be

$$\varepsilon_i = 0 \cdot 0001 x_i$$

This would be inappropriate for variables which approach zero, however, when the choice of ε_i must be made on an absolute rather than a relative basis.

Although the gradient defines the direction of steepest descent, it gives no information concerning the distance of the optimum point. It is not therefore possible to specify an interval over which the one-dimensional search is to be conducted and methods such as the golden-section algorithm cannot be used to determine λ. Fortunately, this one-dimensional search does not need to produce a very accurate result; in fact the use of fairly crude search techniques seems to improve the overall efficiency of the steepest-descent algorithm.

The search method that will be adopted here is repeated doubling. Starting with a value of λ thought to be smaller than the best value, λ is repeatedly doubled. At first the objective function will decrease and then it will start to increase. The value of λ giving the lowest objective function is used as the starting point for the next iteration.

There still remains the problem of selecting a suitable initial value for λ. Since it is likely that the best values of λ on successive steepest-descent iterations will be of similar order of magnitude, the initial value is taken to be the best value of the previous iteration divided by a small constant (say 4). Too large a division ratio will cause unnecessary objective function evaluations, whereas too small a value may result in failure of the iteration to find a better point.

A procedure *steep* for performing a steepest-descent optimization is given below:

182

```
PROCEDURE steep(f: objective; VAR x: vector; n: CARDINAL;
                eps: REAL);
VAR
    g, x0, x1: vector;
    i: CARDINAL;
    lambda, f0, f1, s, q: REAL;
BEGIN
    lambda := 1.0;
    f0 := f(x, n);
    REPEAT
        FOR i := 1 TO n DO
            q := 0.0001*x[i]; x[i] := x[i]+q;
            g[i] := (f(x, n)-f0)/q; x[i] := x[i]-q
        END;
        s := 0.0;
        FOR i := 1 TO n DO s := s+g[i]*g[i] END;
        q := sqrt(s);
        lambda := lambda/8.0;
        f1 := infinity; x0 := x;
        REPEAT
            lambda := 2.0*lambda;
            f0 := f1; x := x1;
            FOR i := 1 TO n DO x1[i] := x0[i]-lambda*g[i]/q END;
            f1 := f(x1, n);
        UNTIL f1 >= f0;
    UNTIL quit()
END steep;
```

The steepest-descent method should be terminated when the distance moved at any iteration is smaller than the specified accuracy. In the procedure *quit* shown below, the distance moved relative to the coordinate value along each axis is compared with the accuracy parameter *eps*. Only when the test succeeds for all n axes is the optimization process terminated. This procedure is local to *steep*.

Experience suggests that steepest-descent operates well during the initial stages of an optimization, while at some distance from the minimum. Scaling of the variables to make the objective function contours approximately spherical will often improve the rate of convergence.

```
PROCEDURE quit(): BOOLEAN;
VAR
    i: CARDINAL;
BEGIN
    FOR i := 1 TO n DO
        IF ABS(lambda*g[i]/(q*x[i])) > eps THEN RETURN FALSE END;
    END;
    RETURN TRUE
END quit;
```

However, close to the minimum where the objective function may exhibit a more complex behaviour, the steepest-descent method tends to be inefficient. In some applications it may be advantageous to use steepest-descent initially, and then switch to an alternative such as simplex for the final optimization stages.

8.15 Least-Squares Optimization

The least-squares optimization method operates on the individual errors and is therefore potentially more efficient than methods which only employ the combined objective function. Least-squares is, however, not of such wide application as the latter methods because its use is confined to minimizing the sum of the squares of the errors. Since it is a gradient method, least-squares optimization is only suitable for optimizing well-behaved error functions.

The starting point for the least-squares method is the sum-of-squares expression for the objective function:

$$f = \sum_{i=1}^{m} e_i^2 \qquad (8.15.1)$$

or in vector notation:

$$f = \mathbf{e}^T \mathbf{e} \qquad (8.15.2)$$

Now the assumption is made that each of the individual errors e_i is linearly dependent on each of the variables x_j. In other words, only the first derivatives in the n-dimensional Taylor expansions are considered to be significant and the higher derivatives are neglected. The values of the errors e_i at a point $\mathbf{x}_0 + \Delta \mathbf{x}$ are given by

$$e_i(\mathbf{x}_0 + \Delta \mathbf{x}) = e_i(\mathbf{x}_0) + \sum_{j=1}^{n} \Delta x_j \frac{\partial e_i}{\partial x_j} \qquad (8.15.3)$$

184

or

$$\mathbf{e} = \mathbf{e}_0 + \mathbf{A} \, \Delta\mathbf{x} \tag{8.15.4}$$

where the $m \times n$ matrix of derivatives $\mathbf{A}$ is given by

$$\mathbf{A} = \begin{bmatrix} \dfrac{\partial e_1}{\partial x_1} & \dfrac{\partial e_1}{\partial x_2} & \cdots & \dfrac{\partial e_1}{\partial x_n} \\[2mm] \dfrac{\partial e_2}{\partial x_1} & & & \\[2mm] \vdots & & & \\[2mm] \dfrac{\partial e_m}{\partial x_1} & & & \dfrac{\partial e_m}{\partial x_n} \end{bmatrix} \tag{8.15.5}$$

and $\mathbf{e}_0$ represents the errors evaluated at the initial point $\mathbf{x}_0$.

At a minimum of the objective function all derivatives of f are zero:

$$\frac{\partial f}{\partial x_j} = 0 \qquad \text{for all } j \tag{8.15.6}$$

or

$$\mathbf{g} = \mathbf{0} \tag{8.15.7}$$

Substituting the expression for the objective function (equation 8.15.1) gives

$$\frac{\partial f}{\partial x_j} = 2 \sum_{i=1}^{m} e_i \frac{\partial e_i}{\partial x_j} = 0 \tag{8.15.8}$$

or

$$\mathbf{g} = 2\mathbf{A}^T\mathbf{e} = \mathbf{0} \tag{8.15.9}$$

Finally, combining equations (8.15.4) and (8.15.9) gives

$$2\mathbf{A}^T\mathbf{e} \quad = 2\mathbf{A}^T\mathbf{e}_0 + 2\mathbf{A}^T\mathbf{A} \, \Delta\mathbf{x} = \mathbf{0}$$
$$\mathbf{A}^T\mathbf{A} \, \Delta\mathbf{x} = -\mathbf{A}^T\mathbf{e}_0 \tag{8.15.10}$$

Solution of this matrix equation should yield a value of $\Delta\mathbf{x}$ for which

$$\mathbf{g}(\mathbf{x}_0 + \Delta\mathbf{x}) = \mathbf{0}$$

so that the point $\mathbf{x}_0 + \Delta\mathbf{x}$ is a minimum. However this equation is only approximate because it was derived on the assumption that the errors vary linearly with the variables around the point $\mathbf{x}_0$. Consequently $\mathbf{x}_0 + \Delta\mathbf{x}$ is only an approximation to the optimum and the process must be repeated until a sufficient degree of accuracy has been attained.

Unfortunately, during the early stages of optimization the least-squares procedure will generate large correction terms $\Delta\mathbf{x}$. The first-order Taylor expansions may then be poor approximations and the optimisation will fail to converge. One way of improving the convergence is to apply damping. A fraction λ of the correction term is used so that the starting point for the next iteration is $\mathbf{x}_0 + \lambda\, \Delta\mathbf{x}$ where λ lies between 0 and 1. Obviously the value of λ must be chosen with some care. Too large a value will result in failure to converge, whereas too small a value will lead to a large number of iterations being necessary. This problem can be overcome by performing a search over the range 0 to 1 for the best value of λ at each iteration.

An alternative approach to overcoming the divergence problems of the basic least-squares method is to introduce an extra term into the equation for $\Delta\mathbf{x}$:

$$[\mathbf{A}^T\mathbf{A} + \lambda\mathbf{I}]\ \Delta\mathbf{x} = -\mathbf{A}^T\mathbf{e}_0 \tag{8.15.11}$$

where λ is an adjustable parameter. As λ tends towards infinity the equation becomes

$$\Delta\mathbf{x} = -\mathbf{A}^T\mathbf{e}_0/\lambda = -\mathbf{g}/2\lambda \tag{8.15.12}$$

This is a step in the steepest descent direction, which is not divergent, and which experience has shown to work well in the early stages of optimization. Small values of λ give a normal least-squares step and therefore the fastest convergence close to the minimum. Varying λ allows the optimization to proceed smoothly from steepest-descent to least-squares form.

A suitable value for λ must be determined at each iteration. Starting from a value of λ that is some fraction (say 1/16) of that used in the previous iteration, a search is performed by repeatedly doubling λ until the objective function evaluated at $\mathbf{x}_0 + \Delta\mathbf{x}$ starts to increase. This search does not involve recalculating the matrix of derivatives. However, for each value of λ the matrix equation must be solved and an evaluation of the objective function performed.

A procedure for performing a least-squares optimization is given below:

```
PROCEDURE lsquares(ev: evaluate; VAR x: vector; n: CARDINAL;
                   eps: REAL);
VAR
    e0, e1, af, dx, x0, x1: vector;
    a, aa, at: matrix;
    i, j, m: CARDINAL;
    lambda, q, f0, f1: REAL;
BEGIN
    lambda := 1.0;
    ev(x, n, e0, m);
    f0 := sumofsquares(e0, m);
    REPEAT
```

```
        FOR i := 1 TO n DO
            q := 0.0001*x[i]; x[i] := x[i]+q;
            ev(x, n, e1, m); x[i] := x[i]-q;
            FOR j := 1 TO m DO a[j,i] := (e1[j]-e0[j])/q END
        END;
        at := a; transpose(at, m, n);
        mxprod(aa, at, a, n, n, m);
        mvprod(af, at, e0, n, m);
        lambda := lambda/32.0;
        f1 := infinity; x0 := x;
        REPEAT
            lambda := 2.0*lambda;
            e0 := e1; f0 := f1; x := x1;
            FOR i := 1 TO n DO aa[i,i] := aa[i,i]+lambda END;
            solveq(aa, dx, af, n);
            FOR i := 1 TO n DO x1[i] := x0[i]-dx[i] END;
            ev(x1, n, e1, m);
            f1 := sumofsquares(e1, m);
        UNTIL f1 >= f0;
    UNTIL quit()
END lsquares;
```

---

evaluate is a procedure type which generates an error vector **e** from a variable vector **x**:

```
TYPE
    objective = PROCEDURE(vector, CARDINAL): REAL;
    evaluate = PROCEDURE(vector, CARDINAL, VAR vector, VAR CARDINAL);
```

---

sumofsquares, as the name suggests, returns the sum of the squares of the m components of the error vector **E**:

```
PROCEDURE sumofsquares(e: vector; m: CARDINAL): REAL;
VAR
    i: CARDINAL;
    q: REAL;
```

```
BEGIN
    q := 0.0;
    FOR i := 1 TO m DO q := q+e[i]*e[i] END;
    RETURN q
END sumofsquares;
```

Termination of the optimization process is controlled by the local procedure *quit*. The criterion used is very similar to that employed for steepest-descent optimization. A *TRUE* result is returned only if correct terms $dx[i]$ are less than $x[i] \times eps$ for all of the variables.

```
PROCEDURE quit(): BOOLEAN;
VAR
    i: CARDINAL;
BEGIN
    FOR i := 1 TO n DO
        IF ABS(dx[i]/x[i]) > eps THEN RETURN FALSE END;
    END;
    RETURN TRUE
END quit;
```

Solution of the matrix equations will yield inaccurate results if the matrix A^TA is ill-conditioned (that is, close to being singular). This situation may arise if the variables $x_1 \ldots x_n$ are of very different magnitudes. If the actual variables cover a wide range of values (greater than, say, 1000 to 1) then scaling should be used to ensure that the optimization variables are of similar magnitude.

Clearly the least-squares optimization method involves considerably more computational effort at each iteration than the simplex or steepest-descent methods. To determine the matrix of derivatives A, the equivalent of $(n + 1)$ evaluations of the objective function must be performed. In addition, the search for a suitable value of λ involves several further objective function evaluations and matrix equation solutions. Nevertheless the convergence rate of the least-squares method (based on the number of objective function evaluations required to achieve a given degree of accuracy in the variables) is superior to that of alternative methods.

Figure 8.14 illustrates the convergence of the simplex, steepest-descent and least-squares methods for a typical optimization problem. Starting

Figure 8.14
Convergence of
a) simplex,
b) steepest-
descent and
c) least-squares
optimization
methods

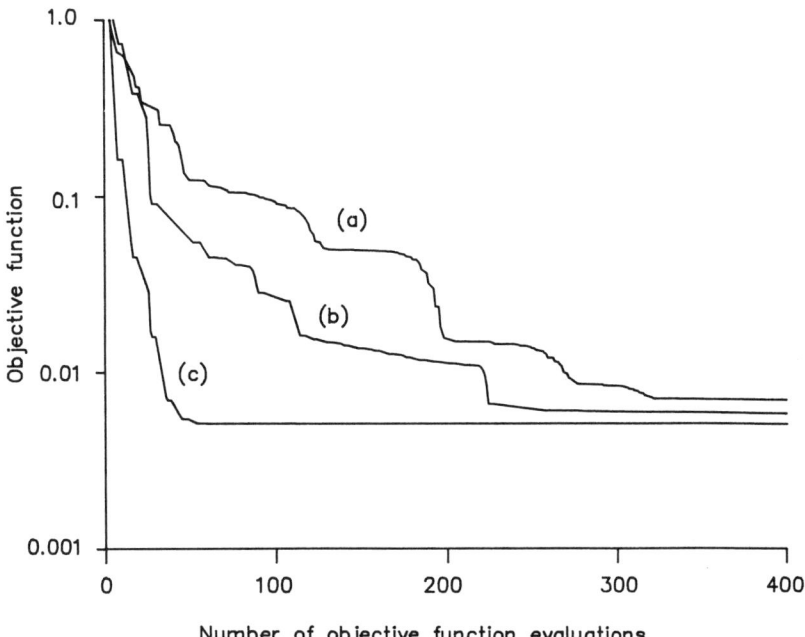

from an initial value of about $1 \cdot 0$, the objective function is reduced to slightly below $0 \cdot 01$. In the case of the simplex and steepest-descent methods this required about 250 objective function evaluations; by contrast the least-squares method needed only 50 evaluations.

Two other points of interest emerge from an inspection of figure 8.14. Different local optima are found by the three optimization methods; however, no general rule concerning which method will find the best local optimum should be deduced from this particular case. Optimization does not proceed smoothly but, particularly in the cases of the simplex and steepest-descent methods, periods of relative stability of the objective function are followed by periods of rapid improvement. This will obviously lead to difficulties in deciding when to terminate the optimization procedure. It is important that this behaviour is borne in mind when selecting a suitable value for the accuracy parameter *eps*.

8.16 Constrained Optimization

Most electronic design problems include constraints on certain of the variables and the application of unconstrained optimization procedures to such problems may lead to these constraints being violated. For example, optimizing the frequency response of a filter network may result in negative values for some of the resistors, capacitors and inductors. There are two approaches which can be used to apply constraints to the variables, while allowing the use of the optimization methods already described.

189

The most obvious way of keeping a variable within a permitted range is to modify the objective function by including a penalty which depends on the value of the variable. Within the permitted range the penalty is small; outside the range the penalty becomes large. Care must be taken in the choice of a suitable penalty function. An abrupt penalty function such as

$$
\begin{aligned}
p_i &= \quad 0 \cdot 0 \quad \text{for} \quad x_i \geqslant 0 \cdot 0 \\
&= 1000 \cdot 0 \quad \text{for} \quad x_i < 0 \cdot 0
\end{aligned}
\tag{8.16.1}
$$

is unsuitable because it is likely to trap the optimization against the hyperplane defined by $x_i = 0$. A better penalty function would be

$$
p_i = \exp(- x_i)
\tag{8.16.2}
$$

because it affects the objective function gradually as x_i approaches zero and becomes negative.

An alternative method for constraining the variables is to use a suitable variable transformation. For example, suppose that the resistor R_i is allowed to have only positive values, and that the corresponding optimisation variable is x_i. The transformation

$$
R_i = x_i^2
\tag{8.16.3}
$$

guarantees that R_i remains positive. Other transformations can limit a component value to within a specified range. For example, the transformation

$$
C_i = \frac{1}{1 + x_i^2}
\tag{8.16.4}
$$

constrains the value of C_i to lie between $0 \cdot 0$ and $1 \cdot 0$. In general the simplest transformations that achieve the required result should be used. Complex transformations produce complicated objective functions which may be difficult to minimize.

8.17 Summary

Computer optimization is a design technique in which network component values are automatically adjusted to obtain the best performance. Optimization is an iterative process. The network is analyzed and an index of performance, the objective function, is evaluated. Then the component values are modified in such a way as to reduce the objective function. After a number of iterations the objective function will show no further improvement, and the process can be terminated. Optimization methods do not guarantee to find the global minimum of the objective function.

The objective function measures the difference between the actual behaviour and the ideal behaviour of a network. In most cases it is a combination of a number of individual differences or errors, and the way that the errors are combined influences the result of optimisation. Sum-of-moduli, maximum-modulus and sum-of-squares are possible methods for combining the errors. Where necessary, individual weightings can be applied to the errors.

Some optimization problems involve only a single variable, and are usually solved by conducting a one-dimensional search over an initial range known to contain an optimum. The golden-section search method is efficient and can be used with discontinuous objective functions.

Optimizing a function of several variables consists of searching the n-dimensional space of the variables to find a minimum of the objective function. All optimization methods are iterative, and fall into one of two classes, namely search methods and gradient methods.

Search methods take no account of the local gradient. The simplex method, which is probably the most generally useful search technique, maintains a solid with $(n + 1)$ vertices (a simplex) in the n-dimensional space. At each iteration the vertex with the highest value of objective function is replaced by a new point with a lower value. Gradually the simplex adapts itself to the local contours of the objective function, and becomes smaller, converging on the optimum.

The simplest of the gradient methods is steepest-descent. In this method the gradient $\mathbf{g}$ of the objective function at some point is evaluated. It is assumed that the optimum lies in the opposite direction to $\mathbf{g}$, so that a one-dimensional search is conducted along the direction $-\mathbf{g}$ to find a minimum of the objective function. This process is repeated until a sufficient degree of accuracy has been attained.

Least-squares is an optimization technique which operates on the individual errors, rather than on the composite objective function. It is potentially more efficient than either simplex or steepest-descent. Each of the errors is expressed as a first-order n-dimensional Taylor expansion. A set of simultaneous linear equations can then be set up, whose solution gives a point where the gradient of the objective function is zero. This corresponds to a minimum of the objective function. In fact, because of the approximation involved in the Taylor expansion, the new point is only an approximation to the optimum and the process must be repeated as necessary.

Simplex, steepest-descent and least-squares optimization techniques place no constraints on the values of the variables. In any practical application, however, the variables must be constrained so that, for example, negative values of capacitance are not generated. One approach to constrained optimization is to apply a penalty to the objective function when the variables exceed their permitted ranges. An alternative approach is to use appropriate transformations between the optimization variables and the component values.

8.18 Problems

1 A third-order control system has a closed-loop transfer function given by

$$H(s) = \frac{48}{60 + 19s + 8s^2 + s^3}$$

The corresponding unit-step response $g(t)$ can be found by inverse Laplace transformation:

$$g(t) = 0 \cdot 8 - 0 \cdot 177 e^{-6 \cdot 50 t} + 0 \cdot 831 e^{-0 \cdot 751 t} \sin(2 \cdot 94 t - 2 \cdot 29)$$

This response overshoots before settling down to a value of $0 \cdot 8$. Use the golden-section method to search for the overshoot peak in the range $t = 0$ to $t = 2$ sec. (Note that the maximum of $g(t)$ is the minimum of $-g(t)$.) Hence determine the percentage overshoot.

2 A third-order low-pass filter has a frequency response function $H(j\omega)$ given by

$$H(j\omega) = \frac{1 + a_1 j\omega + a_2 (j\omega)^2}{1 + b_1 j\omega + b_2 (j\omega)^2 + b_3 (j\omega)^3}$$

Using the simplex method optimize the coefficients of the frequency response function to obtain the best fit to an ideal frequency response $H_0(j\omega)$, where

$$|H_0(j\omega)| = 1 \qquad 0 \leqslant \omega \leqslant 1 \cdot 0$$
$$|H_0(j\omega)| = 0 \qquad \omega \geqslant 1 \cdot 5$$

Repeat this exercise with a factor of four weighting applied to the response for frequencies above $\omega = 1 \cdot 5$.

3 A third-order filter has a unit-step response $g(t)$ given by

$$g(t) = 1 + x_1 \exp(-x_2 t) + x_3 \exp(-x_4 t)\sin(x_5 t + x_6)$$

Using the steepest-descent method optimize the variables $x_1 \ldots x_6$ to obtain the best fit to an ideal step response $g_0(t)$, where

$$g_0(t) = t \qquad 0 \leqslant t < 1$$
$$g_0(t) = 1 \qquad t \geqslant 1$$

Repeat this exercise with a factor of four weighting applied to the response for times greater than $t = 1$.

4 A third-order filter has a frequency response function $H(j\omega)$ of the form:

$$H(j\omega) = \frac{a_0 + a_1 j\omega + a_2 (j\omega)^2 + a_3 (j\omega)^3}{1 + b_1 j\omega + b_2 (j\omega)^2 + b_3 (j\omega)^3}$$

The filter is required to have a gain characteristic $|H_0(j\omega)|$ over the frequency range $\omega = 0\cdot5$ rad/s to $\omega = 2$ rad/s given by

$$|H_0(j\omega)| = \omega^{-1/2}$$

This gain characteristic cannot be obtained exactly from a filter of finite order. Using the least-squares method optimize the coefficients of the third-order frequency response function to obtain the best approximation to the ideal gain characteristic. The three denominator coefficients should be constrained to take only positive values.

5 A pulse-width modulation inverter operates at a frequency of nine times its fundamental sinusoidal output frequency. The output waveform, corresponding to one cycle of the fundamental output frequency, is shown in figure 8.15. Fourier analysis of this waveform gives the frequency components. Cosine components are zero provided that the pulse-width modulation waveform is an odd function around zero:

$$f(-x) = f(2\pi - x) = -f(x)$$

Even harmonics of the fundamental output are zero provided that the waveform is an even function around $\pi/2$:

$$f(\pi - x) = f(x)$$

With these restrictions the four points x_1, x_2, x_3 and x_4 completely define the output waveform, and the amplitude A_n of the nth harmonic, where n is odd, is given by

$$A_n = \frac{8}{n\pi} \{ \tfrac{1}{2} - \cos\, nx_1 + \cos\, nx_2 - \cos\, nx_3 + \cos\, nx_4 \}$$

Using the least-squares method optimize the pulse-width modulation waveform to obtain a fundamental amplitude of $A_1 = 1$, and minimum harmonic components A_3, A_5 and A_7.

Figure 8.15
A pulse-width
modulation
waveform

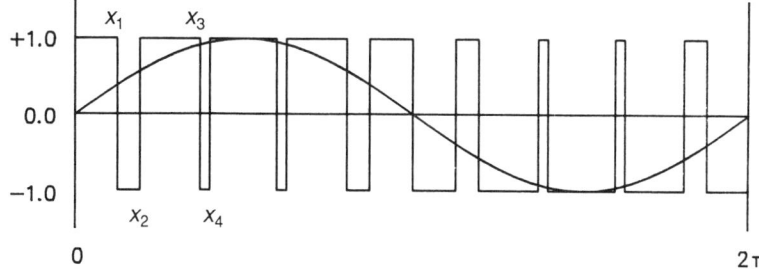

9 D.C. Analysis of Non-linear Networks

9.1 Introduction

All real networks are non-linear. Even a network containing only passive resistors fails to obey superposition when the voltages and currents across individual elements are sufficient to cause significant heating and therefore changes in resistance. Active devices such as bipolar and field-effect transistors are highly non-linear. However, most networks operate over a limited range of voltage and current in regions of near linearity. This is fortunate because it allows linear small-signal models of the non-linear devices to be used, and networks can then be analyzed by linear techniques. Small-signal models are, of course, operating point dependent. D.C. non-linear analysis is therefore performed prior to linear analysis in order to establish the operating points of the devices.

Consider the unsaturated characteristic of a forward-biased bipolar transistor:

$$I_C = I_S \exp(V_{BE}/V_T) \tag{9.1.1}$$

where I_C is the collector current, V_{BE} is the base–emitter voltage, I_S is a constant and $V_T = kT/q$ is the thermal voltage. This expression can be differentiated to give

$$\frac{\mathrm{d}I_C}{\mathrm{d}V_{BE}} = \frac{I_S}{V_T} \exp(V_{BE}/V_T) = \frac{I_C}{V_T} \tag{9.1.2}$$

A small change δV_{BE} in base–emitter voltage will lead to a proportional change $\delta I_C = \delta V_{BE}(I_C/V_T)$ in collector current. In other words, for small signals the bipolar transistor can be regarded as a linear voltage-controlled current source of transconductance I_C/V_T. Before any linear analysis method can be used, however, the value of the transconductance must be determined, and this depends on the bias current I_C.

Most active networks contain a number of non-linear devices whose voltages and currents are mutually dependent and determination of the bias points involves the solution of simultaneous non-linear equations.

Fortunately only a d.c. solution is required so that capacitors can be replaced by open circuits and inductors by short circuits.

In principle the non-linear equations may have several solutions, representing different stable d.c. states of the network, and there can be no guarantee that all of these solutions will be found. In practice, d.c. non-linear analysis is useful for determining the normal operating point of a network, but is less successful in looking for unwanted latch-up states.

The case of networks containing a single non-linear element will be considered first; then the techniques used for analyzing such networks will be generalized to deal with several non-linear elements.

9.2 Networks Containing a Single Non-linear Element

The network shown in figure 9.1 represents a bipolar transistor amplifier with parallel voltage feedback. Neglecting the base current of the transistor leads to the following d.c. network equations:

$$\begin{aligned} I_1 &= I_S \exp(V_1/V_T) \\ V_1 &= R_3 I_2 \\ V_2 &= V_1 + R_2 I_2 \\ V_2 &= V_S - R_1(I_1 + I_2) \end{aligned} \tag{9.2.1}$$

Eliminating V_1, I_1 and I_2 gives a single equation for V_2:

$$V_2 + \frac{V_2 R_1}{(R_2 + R_3)} + R_1 I_S \exp\left\{\frac{V_2 R_3}{V_T(R_2 + R_3)}\right\} - V_S = 0 \tag{9.2.2}$$

or

$$f(V_2) = 0 \tag{9.2.3}$$

One approach to solving this equation is to employ a binary search strategy.

Figure 9.1
A bipolar
transistor
amplifier

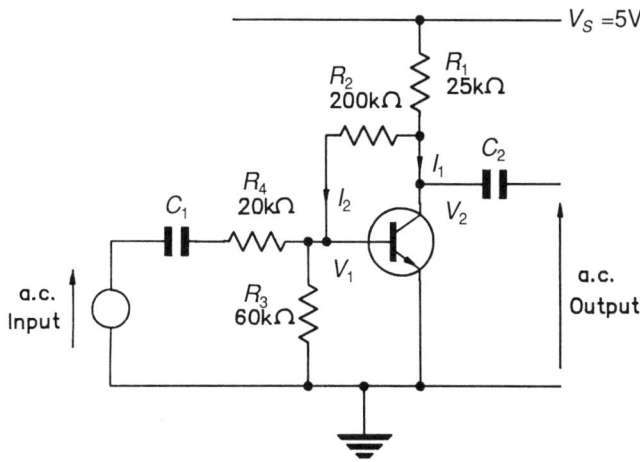

195

If an interval a to b can be found for which $f(a)$ and $f(b)$ have different signs, then at least one root of the equation must lie between a and b. The root can be localized by successively bisecting the interval. Let c represent the midpoint of the interval a to b. If $f(a) \times f(c) < 0$ then the root lies between a and c; otherwise it must lie between c and b. This process is repeated and at each stage the size of the interval is halved. Eventually the interval becomes less than the required accuracy of the root, and the search is terminated. Figure 9.2 illustrates a binary search performed on a function with a single root in the initial interval a_1 to b_1.

A procedure *binarysearch* which finds a root of the equation $f(x) = 0$ in the interval a to b, to an absolute accuracy *eps*, is given below:

```
PROCEDURE binarysearch(f: function; a, b: REAL; eps: REAL): REAL;
VAR
     xa, xb, xc, fa, fb, fc: REAL;
BEGIN
     xa := a; fa := f(xa);
     xb := b; fb := f(xb);
     IF fa*fb > 0.0 THEN
          WriteString('No solution');
          WriteLn;
          HALT
     END;
     REPEAT
          xc := 0.5*(xa+xb); fc := f(xc);
          IF fa*fc < 0.0 THEN
               xb := xc; fb := fc
          ELSE
               xa := xc; fa := fc
          END;
     UNTIL ABS(xb-xa) < eps;
     RETURN xc
END binarysearch;
```

This procedure was used to obtain the solution of equation (9.2.2), with transistor parameters $I_S = 1 \times 10^{-13}$ A and $V_T = 26$ mV. Starting from an initial search interval 0 V to 5 V, the solution $V_2 = 2 \cdot 332245$ V was found to an accuracy of 1 μV in 22 iterations.

Binary search has the advantages of simplicity and a constant

Figure 9.2
Solution by
binary search

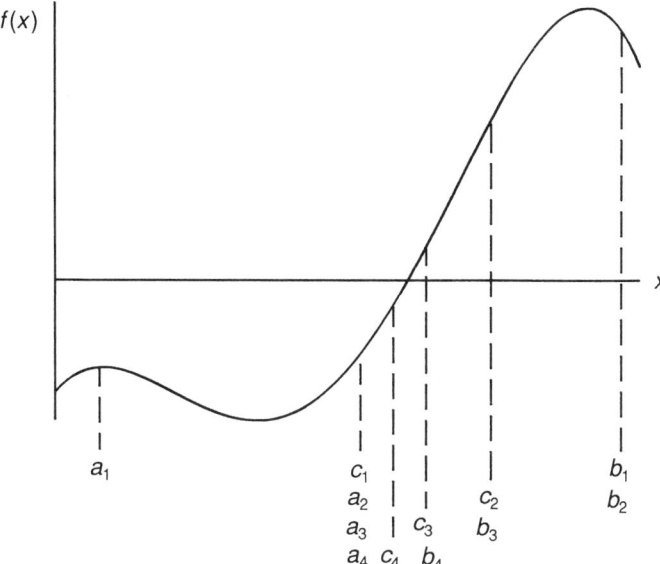

convergence rate. In fact, the number of iterations n necessary to achieve an accuracy ε can be determined in advance:

$$n = \log_2\{(b - a)/\varepsilon\} \qquad (9.2.4)$$

Nevertheless binary search can prove unsatisfactory for use with some functions as illustrated in figure 9.3. This function has two adjacent roots and for a binary search to be successful one end of the initial interval must lie between the roots. Since the roots are not known in advance it is unlikely that this condition will be met.

An extreme example of this problem is where the roots are actually coincident as illustrated in figure 9.4. Binary search will always fail in this case. Fortunately, in practice, non-linear network equations are unlikely to have coincident roots.

An alternative approach to solving non-linear equations in a single variable is the Newton–Raphson method. In chapter 3 the Newton–Raphson method, and its use in determining the complex poles and

Figure 9.3
A function with
adjacent roots

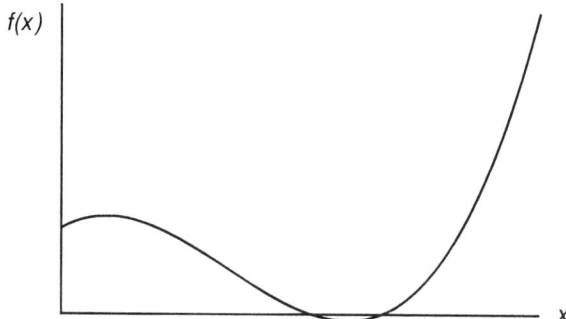

197

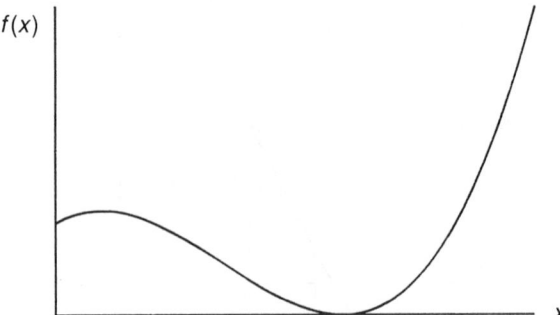

Figure 9.4
A function with coincident roots

zeros of the transfer function, were discussed in detail. The derivation of the Newton–Raphson formula will not be repeated here. Instead the result will simply be quoted:

$$x_{i+1} = x_i - \{f(x_i)/f'(x_i)\} \qquad (9.2.5)$$

Given a point x_i, this formula generates a new point x_{i+1} which should be a better approximation to the root; it can be applied iteratively until the required degree of accuracy is obtained.

When used to solve non-linear network equations the variables and functions are of course, real. The procedure *newton* given below finds a root of the equation $f(x) = 0$ to an absolute accuracy *eps* starting from an initial point x_0:

```
PROCEDURE newton(f: function; x0: REAL; eps: REAL): REAL;
VAR
    x, fx, dfdx, dx: REAL;
BEGIN
    x := x0;
    REPEAT
        fx := f(x); dx := 0.0001*x;
        dfdx := (f(x+dx)-fx)/dx;
        dx := -fx/dfdx;
        x := x+dx;
    UNTIL ABS(dx) < eps;
    RETURN x
END newton;
```

Close to a root the convergence of the Newton–Raphson method is particularly rapid. On the other hand during the initial stages, while remote from the root, the convergence may be quite slow. For example, when the procedure *newton* was used to solve equation (9.2.2) starting from $x_0 = 2$ V,

198

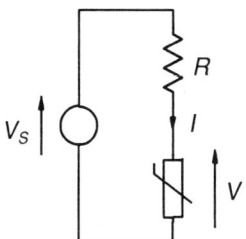

Figure 9.5
A non-linear
potential
divider

a total of 14 iterations were required to obtain an accuracy of $1\ \mu$V. Since each iteration involves two function evaluations, the procedure *newton* is less efficient than *binarysearch* for solving the equation to this level of accuracy. There is, in fact, no guarantee that the Newton–Raphson method will converge at all, although in practice it is nearly always successful in finding the solution to non-linear network equations.

An advantage of the Newton–Raphson method over binary search is that only an initial approximation to the root is required, rather than an interval known to contain the root. It is usually successful in finding one of a pair of adjacent roots, although for reasons discussed in chapter 3 it may fail if the roots are coincident.

Not all solutions to non-linear network equations are necessarily physically realistic. Consider the non-linear potential divider shown in figure 9.5. The current I must satisfy the characteristic of the non-linear element:

$$I = g(V) \qquad\qquad (9.2.6)$$

and must also satisfy Ohm's law for the linear resistor:

$$I = (V_S - V)/R \qquad\qquad (9.2.7)$$

Now suppose that the characteristic $g(V)$ of the non-linear element is similar to that shown in figure 9.6. A tunnel diode is an example of a device with such a characteristic. Points of intersection between this non-linear characteristic and a line representing equation (9.2.7) give the values of V and I which satisfy both equations. Alternatively the equations can be

Figure 9.6
A non-linear
characteristic

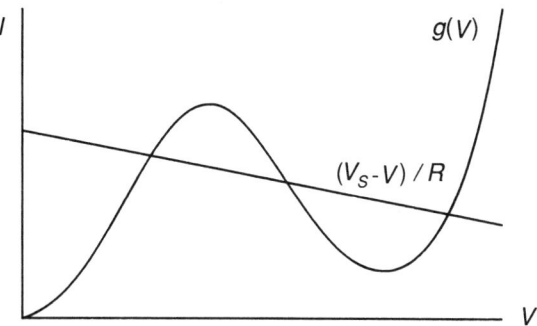

199

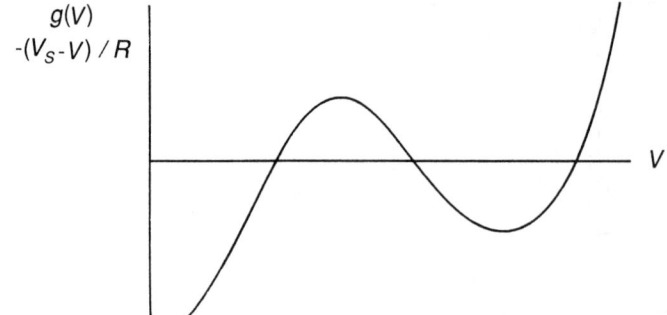

Figure 9.7
Solutions of
the non-linear
equation

$g(V)$
$-(V_S-V)/R$

V

combined algebraically, and the current I eliminated, to give

$$g(V) - (V_S - V)/R = 0 \qquad (9.2.8)$$

Figure 9.7 shows the variation of the left-hand side of this equation with the variable V.

There are three solutions to the equation corresponding to the points of intersection in figure 9.6. All of these solutions can be found by the use of either binary search or the Newton–Raphson method. However, only two of the solutions represent stable d.c. bias points of the network; the middle point is in fact unstable. This example serves as a warning that care must be taken when interpreting the solutions of non-linear network equations.

9.3 General Non-linear Network Equations

The tunnel diode characteristic shown in figure 9.6 is an example of a voltage-controlled non-linearity which can be written in the form:

$$I = f(V) \qquad (9.3.1)$$

where the current I is uniquely defined by the voltage V. It is however possible to have a characteristic which cannot be expressed in this way, as illustrated in figure 9.8.

For a range of voltages there are three possible currents. This

Figure 9.8
A current-
controlled non-
linear
characteristic

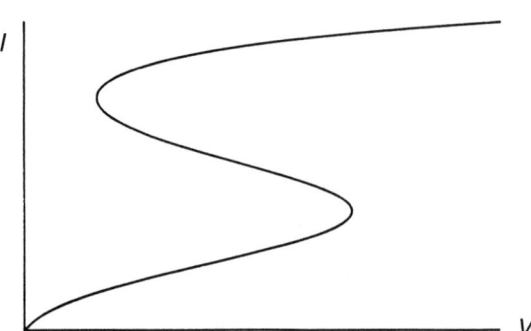

I

V

characteristic must therefore be expressed in current-controlled form:

$$V = f(I) \tag{9.3.2}$$

Many devices have characteristics which can be expressed in either way. For example, the current I in a forward-biased junction diode is given by

$$I = I_S \exp(V/mV_T) \tag{9.3.3}$$

where V is the voltage across the diode and m is a constant which lies between 1 and 2. This equation can be rewritten as

$$V = mV_T \ln (I/I_S) \tag{9.3.4}$$

For the purposes of non-linear network analysis, therefore, the diode can be considered to be either voltage- or current-controlled.

In the case of three terminal devices such as bipolar transistors the characteristic may depend on more than one independent variable. For example, equation (9.1.1) represents the unsaturated characteristic ($V_{BC} < 0$ V) of a bipolar transistor. Close to saturation the base-collector voltage V_{BC} has a significant effect on the collector current and can no longer be neglected. Thus

$$I_C = f(V_{BE}, V_{BC}) \tag{9.3.5}$$

The characteristic of a field-effect transistor can be expressed in a similar way:

$$I_D = f(V_{GS}, V_{GD}) \tag{9.3.6}$$

For the rest of this discussion it will be assumed that the currents in all non-linear devices are voltage controlled, with the device current being dependent on one or more voltages:

$$I = f(V_1, V_2 \dots V_n) \tag{9.3.7}$$

Consider now a network consisting of such voltage-controlled non-linear elements. This does not, of course, exclude resistors and other devices normally considered to be linear because their characteristics are simply special cases of equation (9.3.7). Kirchhoff's current law can be applied to each node of the network, excluding the reference node, so that at node i,

$$f_{i0}(V_1, V_2 \dots V_n) + f_{i1}(V_1, V_2 \dots V_n) + f_{i2}(V_1, V_2 \dots V_n) + \dots = 0 \tag{9.3.8}$$

where f_{ij} represents a non-linear voltage-controlled current flowing from node j to node i. In principle each of the currents can depend on all of the nodal voltages $V_1, V_2 \dots V_n$. If there are n independent nodal voltages, then there will be n non-linear equations similar to equation (9.3.8):

$$f_{10}(V_1 \ldots V_n) + \qquad\qquad f_{12}(V_1 \ldots V_n) + \cdots + f_{1n}(V_1 \ldots V_n) = 0$$
$$f_{20}(V_1 \ldots V_n) + f_{21}(V_1 \ldots V_n) + \qquad\qquad + \cdots + f_{2n}(V_1 \ldots V_n) = 0$$
$$\vdots$$
$$f_{n0}(V_1 \ldots V_n) + f_{n1}(V_1 \ldots V_n) + f_{n2}(V_1 \ldots V_n) + \cdots \qquad\qquad = 0$$

$$(9.3.9)$$

In practice most of these voltage-controlled currents will be zero. The non-linear functions f_{ij} in any equation can be combined to give

$$f_1(V_1, V_2 \ldots V_n) = 0$$
$$f_2(V_1, V_2 \ldots V_n) = 0 \qquad\qquad (9.3.10)$$
$$\vdots$$
$$f_n(V_1, V_2 \ldots V_n) = 0$$

Finally these simultaneous non-linear equations can be expressed in vector form:

$$\mathbf{f}(\mathbf{x}) = \mathbf{0} \qquad\qquad (9.3.11)$$

where

$$\mathbf{f}(\mathbf{x}) = \begin{bmatrix} f_1(\mathbf{x}) \\ f_2(\mathbf{x}) \\ \vdots \\ f_n(\mathbf{x}) \end{bmatrix} \qquad\qquad (9.3.12)$$

and

$$\mathbf{x} = \begin{bmatrix} V_1 \\ V_2 \\ \vdots \\ V_n \end{bmatrix} \qquad\qquad (9.3.13)$$

Consider the two-transistor a.c. voltage amplifier network shown in figure 9.9. To simplify the d.c. analysis any capacitors are replaced by open circuits, and any inductors are replaced by short circuits. The resulting network is shown in figure 9.10.

The supply rail voltage V_S is fixed relative to ground and Kirchhoff's current law cannot be applied to this node. This leaves four independent nodes whose voltages are determined by four simultaneous non-linear equations:

$$(V_S - V_1)/R_1 - V_1/R_2 = 0$$
$$I_S \exp\{(V_1 - V_2)/V_T\} + (V_4 - V_2)/R_5 = 0$$
$$(V_S - V_3)/R_3 - I_S \exp\{(V_1 - V_2)/V_T\} = 0$$
$$I_S \exp\{(V_S - V_3)/V_T\} + (V_2 - V_4)/R_5 - V_4/R_6 = 0$$

In the following section a method will be described for obtaining a solution to simultaneous non-linear equations.

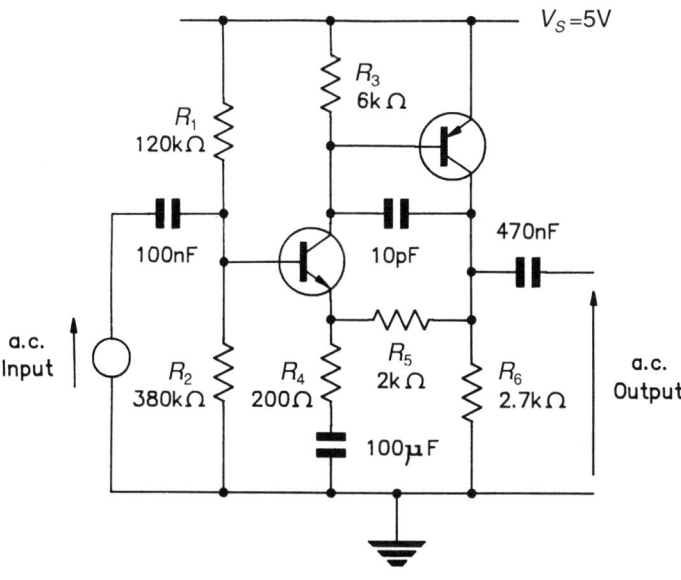

Figure 9.9
A two-transistor a.c. voltage amplifier

$V_S = 5V$

R_3
$6k\,\Omega$

R_1
$120k\Omega$

$100nF$

$10pF$

$470nF$

a.c. Input

R_2
$380k\Omega$

R_4
200Ω

R_5
$2k\,\Omega$

R_6
$2.7k\Omega$

a.c. Output

$100\mu F$

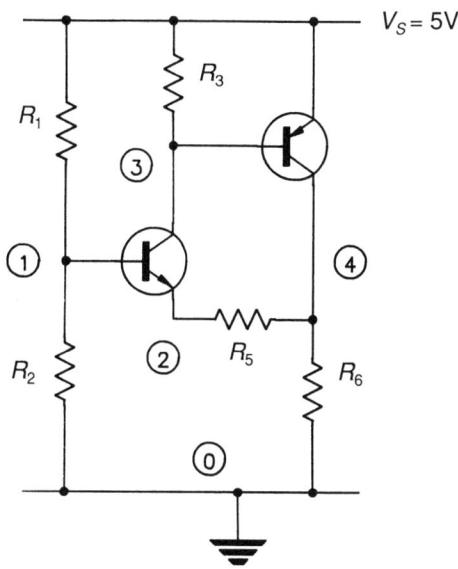

Figure 9.10
D.C. elements of the voltage amplifier

$V_S = 5V$

R_3

R_1

③

①

④

R_2

②

R_5

R_6

⓪

9.4 Multidimensional Newton–Raphson Algorithm

A solution is required to the set of n simultaneous non-linear equations:

$$\mathbf{f}(\mathbf{x}) = \mathbf{0} \tag{9.4.1}$$

The method that will be described is a generalization of the Newton–Raphson technique. Suppose that $\mathbf{x}$ is an initial approximation to a solution of equation (9.4.1). Each of the component functions f_i of $\mathbf{f}$ can

203

be expanded as a Taylor series around this point:

$$f_i(\mathbf{x} + \mathbf{h}) = f_i(\mathbf{x}) + \sum_{j=1}^{n} h_j \frac{\partial f_i}{\partial x_j} + \cdots \tag{9.4.2}$$

where

$$\mathbf{h} = \begin{bmatrix} h_1 \\ h_2 \\ \vdots \\ h_n \end{bmatrix}$$

If $\mathbf{x} + \mathbf{h}$ is a solution of this equation, then

$$f_i(\mathbf{x} + \mathbf{h}) = 0 \tag{9.4.3}$$

Provided that the initial estimate $\mathbf{x}$ is close to the solution, then $\mathbf{h}$ will be small and terms involving second and higher derivatives in the Taylor expansion can be neglected. Thus

$$f_i(\mathbf{x}) + \sum_{j=1}^{n} h_j \frac{\partial f_i}{\partial x_j} = 0 \tag{9.4.4}$$

An equation similar to this exists for each of the component functions. The set of equations can be written in vector-matrix form:

$$\mathbf{f}(\mathbf{x}) + \mathbf{A}(\mathbf{x})\mathbf{h} = \mathbf{0} \tag{9.4.5}$$

where $\mathbf{A}(\mathbf{x})$ is the $n \times n$ matrix of partial derivatives:

$$\mathbf{A}(\mathbf{x}) = \begin{bmatrix} \dfrac{\partial f_1}{\partial x_1} & \dfrac{\partial f_1}{\partial x_2} & \cdots & \dfrac{\partial f_1}{\partial x_n} \\[2mm] \dfrac{\partial f_2}{\partial x_1} & & & \\[2mm] \vdots & & & \\[2mm] \dfrac{\partial f_n}{\partial x_1} & & & \dfrac{\partial f_n}{\partial x_n} \end{bmatrix} \tag{9.4.6}$$

Equation (9.4.5) can be written in the form:

$$\mathbf{A}(\mathbf{x})\mathbf{h} = -\mathbf{f}(\mathbf{x}) \tag{9.4.7}$$

Solving this set of simultaneous linear equations using, for example, Gaussian elimination, gives the values of $\mathbf{h}$. In fact, the point $\mathbf{x} + \mathbf{h}$ will not be an exact solution of equation (9.4.1) because of the approximation involved in neglecting the higher partial derivatives in the Taylor expansion. The process must be repeated until a solution of sufficient accuracy is obtained. If $\mathbf{x}_i$ is the value of $\mathbf{x}$ at the ith iteration, then a better approximation to the solution is given by

$$\mathbf{x}_{i+1} = \mathbf{x}_i - [\mathbf{A}(\mathbf{x}_i)]^{-1}\mathbf{f}(\mathbf{x}_i) \tag{9.4.8}$$

This is the multidimensional Newton–Raphson iteration formula.

In some cases it may be possible to obtain analytically the partial derivatives which make up $\mathbf{A}$. A general-purpose routine for solving non-linear equations should, however, not rely on the analytical derivatives being available. Instead the derivatives should be determined numerically:

$$\frac{\partial f_i}{\partial x_j} \approx \frac{f_i(x_1, \ldots x_j + \varepsilon_j, \ldots x_n) - f_i(x_1, \ldots x_j, \ldots x_n)}{\varepsilon_j} \qquad (9.4.9)$$

Calculation of all of the elements of $\mathbf{A}$ involves $n + 1$ evaluations of each of the n functions f_i. The increments ε_j must be chosen carefully. Large values of ε_j will lead to the formula given above being a poor approximation because of the effect of neglecting the higher derivatives. On the other hand if ε_j is too small then subtraction of nearly equal quantities also gives a poor accuracy. Provided that the variable x_j does not approach zero, a suitable choice for ε_j might be

$$\varepsilon_j = 0 \cdot 0001 x_j$$

The procedure *simnonlin* shown below solves a set of non-linear equations to an absolute accuracy *eps*:

```
PROCEDURE simnonlin(fn: equations; VAR x: vector; n: CARDINAL;
                eps: REAL);

VAR
    h, f0, f1: vector;
    a: matrix;
    i, j: CARDINAL;
    q: REAL;


PROCEDURE quit(): BOOLEAN;
VAR
    i: CARDINAL;
BEGIN
    FOR i := 1 TO n DO
        IF ABS(h[i]) > eps THEN RETURN FALSE END;
    END;
    RETURN TRUE
END quit;
```

```
BEGIN
    REPEAT
        fn(x, f0);
        FOR i := 1 TO n DO
            q := 0.0001*x[i]; x[i] := x[i]+q;
            fn(x, f1); x[i] := x[i]-q;
            FOR j := 1 TO n DO a[j,i] := (f1[j]-f0[j])/q END
        END;
        solveq(a, h, f0, n);
        FOR i := 1 TO n DO x[i] := x[i]-h[i] END;
    UNTIL quit()
END simnonlin;
```

The iterations are terminated when all the corrections h_i are less than the required accuracy, as determined by the local procedure *quit*. *simnonlin* makes use of a procedure type *equations* which is defined as follows:

```
TYPE
    index = [0..maxsize];
    vector = ARRAY index OF REAL;
    equations = PROCEDURE(vector, VAR vector);
```

A procedure *fn* of type *equations* must be provided which, given a vector of variables x_i, generates the corresponding vector of functions f_j. For example the procedure shown below could be used to set up the non-linear network equations for the voltage amplifier discussed in the previous section.

```
PROCEDURE fn(x: vector; VAR f: vector);
CONST
    Is = 1.0E-13;
    Vt = 26.0E-3;
    Vs = 5.0;
    R1 = 120.0E3;   R2 = 380.0E3;   R3 = 6.0E3;
    R4 = 200.0;     R5 = 2.0E3;     R6 = 2.7E3;
VAR
    V1, V2, V3, V4: REAL;
```

```
BEGIN
    V1 := x[1]; V2 := x[2]; V3 := x[3]; V4 := x[4];
    f[1] := (Vs-V1)/R1-V1/R2;
    f[2] := Is*exp((V1-V2)/Vt)+(V4-V2)/R5;
    f[3] := (Vs-V3)/R3-Is*exp((V1-V2)/Vt);
    f[4] := Is*exp((Vs-V3)/Vt)+(V2-V4)/R5-V4/R6;
END fn;
```

When *simnonlin* was used to find the solution of these equations to an accuracy of 10^{-8}, the following results were obtained after 22 iterations:

$$V_1 = 3 \cdot 8000000000\text{E} + 000$$
$$V_2 = 3 \cdot 2612150577\text{E} + 000$$
$$V_3 = 4 \cdot 4004606587\text{E} + 000$$
$$V_4 = 3 \cdot 0613686106\text{E} + 000$$

The Newton–Raphson method may fail if the matrix **A** is poorly conditioned (that is nearly singular) because in this case the solution to equation (9.4.7) is likely to be inaccurate or even meaningless. Poor conditioning of **A** can result from the variables x_i being of widely different magnitudes. For example, a network might contain voltages of the order 1 V, and currents of the order 10^{-6} A. If the variables were to consist of a mixture of such voltages and currents, then there would certainly be problems. The obvious solution is to scale the variables: current variables are multiplied by some scaling factor k to bring their magnitudes in line with the voltage variables. If necessary each of the variables can be given an individual scaling factor.

Whilst there is no guarantee that the Newton–Raphson method will converge (and some reasons for failure to converge have been discussed in chapter 3) it is in practice usually successful in solving non-linear network equations if the starting point is reasonable. Devices such as bipolar transistors which have exponential characteristics should always be given an initial voltage within their normal operating range, otherwise wild variations in the currents may lead to numerical overflow. For example, a bipolar transistor with an initial value of $V_{BE} = 4$ V would have a theoretical collector current of around 10^{54} A. This can be avoided by setting all the base–emitter voltages to $0 \cdot 5$ V initially.

The main drawback of the multidimensional Newton–Raphson method is the need, at each iteration, to evaluate the matrix of partial derivatives **A**. Several quasi-Newton methods have been developed which aim to update **A** at each stage without using extra function evaluations to obtain the partial derivatives. These methods are considerably more complex than the Newton–Raphson method and tend to show a significant improvement in efficiency only with certain classes of equations.

9.5 Component Value Determination

A common problem in electronics which involves simultaneous non-linear equations is the selection of component values to generate a desired frequency response function. Consider, for example, the singly-terminated passive ladder filter shown in figure 9.11.

The components are to be chosen to give a third-order elliptic response with a cut-off frequency of 1 kHz, 3 dB passband ripple and 30 dB stopband attenuation:

$$H(j\omega) = \frac{1 + 9 \cdot 56 \times 10^{-9}(j\omega)^2}{1 + (4 \cdot 93 \times 10^{-4}j\omega) + (4 \cdot 75 \times 10^{-8}(j\omega)^2) + (1 \cdot 29 \times 10^{-11}(j\omega)^3)}$$

Symbolic analysis of the network, either manually or by computer, gives expressions for each of the coefficients of $j\omega$ in the frequency response function:

Numerator:

$+ R_0$

$+ R_0 L_3 C_1 (j\omega)^2$

Denominator:

$+ R_0$

$+ L_2(j\omega) + L_1(j\omega)$

$+ R_0 L_3 C_1(j\omega)^2 + R_0 L_1 C_1(j\omega)^2$

$+ L_2 L_3 C_1(j\omega)^3 + L_1 L_3 C_1(j\omega)^3 + L_1 L_2 C_1(j\omega)^3$

These symbolic numerator and denominator coefficients are first divided by R_0 to make the zero-order terms unity:

Numerator:

$+ 1$

$+ L_3 C_1 (j\omega)^2$

Figure 9.11
Third-order
low-pass
singly-
terminated
ladder filter

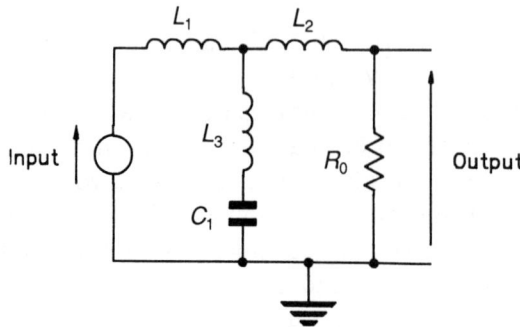

Denominator:

$$+1$$
$$+ L_2/R_0(j\omega) + L_1/R_0(j\omega)$$
$$+ L_3C_1(j\omega)^2 + L_1C_1(j\omega)^2$$
$$+ L_2L_3C_1/R_0(j\omega)^3 + L_1L_3C_1/R_0(j\omega)^3 + L_1L_2C_1/R_0(j\omega)^3$$

Then the symbolic coefficients are equated to their required values giving four simultaneous non-linear equations in five unknowns. One of the component values must therefore be chosen arbitrarily, thereby defining the impedance of the network:

$$R_0 = 1 \text{ k}\Omega$$

Because of the wide range of component values that may be found in a typical network, ranging from capacitors of 10^{-12} F to resistors of $10^6 \Omega$, scaling of the variables will almost certainly be necessary. It is convenient to use scaling factors which make all of the variables x_i dimensionless and of magnitude close to unity:

$$\omega_0 L_1/R_0 = x_1$$
$$\omega_0 L_2/R_0 = x_2$$
$$\omega_0 L_3/R_0 = x_3$$
$$\omega_0 C_1 R_0 = x_4$$

where ω_0 is the approximate cut-off frequency of the filter. The equations can then be solved to give

$$L_1 = 2\cdot78 \times 10^{-1} \text{ H}$$
$$L_2 = 2\cdot15 \times 10^{-1} \text{ H}$$
$$L_3 = 7\cdot01 \times 10^{-2} \text{ H}$$
$$C_1 = 1\cdot36 \times 10^{-7} \text{ F}$$

Since all the variables have been scaled to a magnitude of around unity, the Newton–Raphson iterations can be started from the point $x_i = 1$.

9.6 Summary

D.C. non-linear analysis is used to establish the operating points, and therefore the linear small-signal models, of the non-linear devices in a network. In general d.c. analysis of a non-linear network will lead to a set of simultaneous non-linear equations which may have several solutions representing different stable d.c. states of the network. Some solutions may correspond to unstable states.

Simple networks containing a single non-linear element require only a single equation to be solved. If an interval of the variable can be found which contains a root, then repeated bisection of the interval allows the root to be localized to any desired degree of accuracy. Repeated bisection is not suitable for solving equations with adjacent roots and will always fail if two roots are coincident.

An alternative to repeated bisection is the Newton–Raphson method. This has the advantage of very rapid convergence close to a root, and of only requiring an initial estimate, rather than an initial search interval. Although convergence of the Newton–Raphson algorithm is not guaranteed, it is usually successful in solving non-linear network equations.

Most non-linear devices are voltage controlled, that is the current in the device is a single-valued function of the device voltages. A network containing such devices can be analyzed by applying Kirchhoff's current law to each of the n nodes of the network. This leads to a set of n simultaneous non-linear equations in the n nodal voltages.

The Newton–Raphson algorithm can be generalized to deal with simultaneous equations. As in its original form, the multidimensional Newton–Raphson algorithm is an iterative technique with the solution being refined at each stage. Convergence is likely to be poor if the variables are of widely different magnitudes, and scaling may be necessary. The multidimensional Newton–Raphson algorithm is fairly demanding computationally, involving the numerical evaluation of n^2 partial derivatives at each iteration.

Another application of the multidimensional Newton–Raphson method is in the calculation of component values of a network to give a desired frequency response function. Comparing coefficients of $j\omega$ in the symbolic frequency response function of the network and in the desired frequency response function leads to a set of simultaneous non-linear equations.

9.7 Problems

1 The network shown in figure 9.12 represents a potential divider incorporating a thermistor. Thermistors have a resistance R_{th} which varies with absolute temperature T according to

$$R_{th} = \frac{V}{I} = A \, \exp\left(\frac{B}{T}\right)$$

Figure 9.12
A non-linear
potential
divider

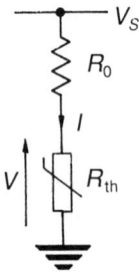

210

where A and B are constants characteristic of particular thermistors. Power dissipation in the thermistor will raise its temperature T above the ambient temperature T_0:

$$T = T_0 + DVI$$

where D is the thermal resistance to ambient of the device. These equations can be combined, with the elimination of T, to give the $V-I$ characteristic:

$$\frac{V}{I} = A \, \exp\left(\frac{B}{T_0 + DVI}\right)$$

A miniature bead thermistor has the following properties:

$$A = 0 \cdot 08 \; \Omega \qquad B = 4000 \; K \qquad D = 1400°K/W$$

Taking the ambient temperature to be $T_0 = 298$ K, solve the $V-I$ equation by binary search to obtain V for values of I between 0 and 20 mA. Observe that the thermistor is a current-controlled non-linear element.

The potential divider has a supply voltage of $V_S = 15$ V, and a linear resistor of value $R_0 = 300 \; \Omega$. Show that there are three values of V and I which satisfy Ohm's law for the resistor:

$$V = V_S - IR_0$$

and the $V-I$ equation for the thermistor. (Only two of these values represent stable operating points.)

2 Figure 9.13 shows one section of a piecewise-linear shaping network. In a practical shaping network the current shown flowing to ground would flow into the virtual ground of an operational amplifier. The diodes have a voltage-current characteristic given by

$$I = I_S \exp(V/mV_T)$$

where $I_S = 10^{-13}$ A, $m = 1 \cdot 5$ and $V_T = 26$ mV. Using this characteristic it

Figure 9.13
A piecewise-
linear section

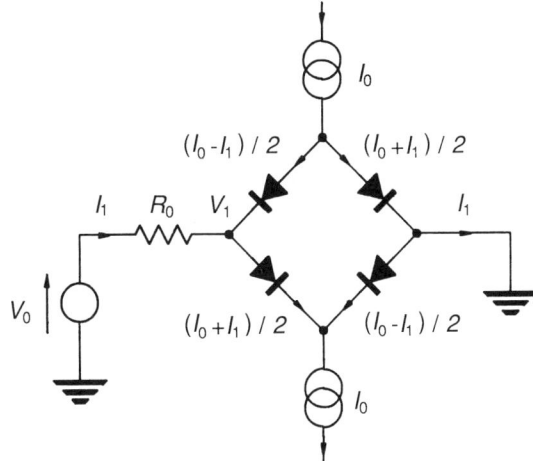

can easily be shown that the input voltage V_0 and the output current I_1 are related by

$$V_0 - R_0 I_1 = mV_T \ln\left\{\frac{I_0 + I_1}{I_0 - I_1}\right\}$$

Given that $I_0 = 100\ \mu A$ and $R_0 = 10\ k\Omega$, solve this equation by binary search to obtain I_1 for values of V_0 ranging from $-1\cdot5$ V to $1\cdot5$ V. To prevent numerical overflow use an initial search interval of

$$-99\cdot99999\ \mu A\ \text{to}\ 99\cdot99999\ \mu A$$

3 The network shown in figure 9.14 represents an a.c. voltage amplifier with overall negative feedback. The three bipolar transistors have identical characteristics:

$$I_C = I_S \exp(V_{BE}/V_T)$$

where $I_S = 10^{-13}$ A and $V_T = 26$ mV. Base currents may be neglected.

Using the multidimensional Newton–Raphson algorithm, determine the operating points (that is the base–emitter voltages) of the bipolar transistors, and the d.c. level of the output.

Figure 9.14
An a.c. voltage
amplifier

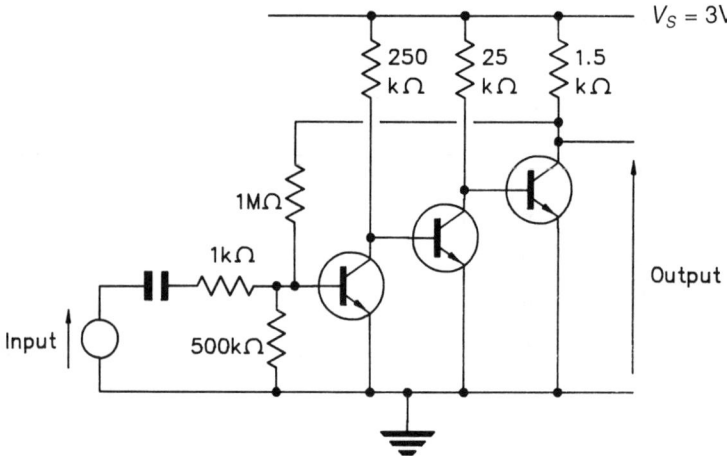

4 Figure 9.15 shows the network diagram of an a.c. voltage amplifier incorporating a junction field-effect transistor and a bipolar transistor. The characteristics of the bipolar and field-effect transistors are given respectively by

$$I_C = I_S \exp(V_{BE}/V_T)$$
$$I_D = I_{DSS}(1 - V_{GS}/V_p)^2$$

where $I_S = 10^{-12}$ A, $V_T = 26$ mV, $I_{DSS} = 5$ mA and $V_p = -2$ V. The base current of the bipolar transistor may be neglected.

Use the multidimensional Newton–Raphson algorithm to determine the operating points of the transistors and the d.c. level of the output.

Figure 9.15
An a.c. voltage
amplifier

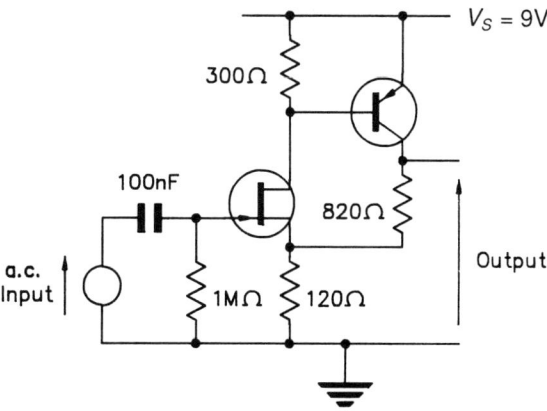

$V_S = 9V$

300Ω

100nF

820Ω

a.c.
Input

Output

1MΩ   120Ω

Figure 9.16
A singly-
terminated
ladder filter

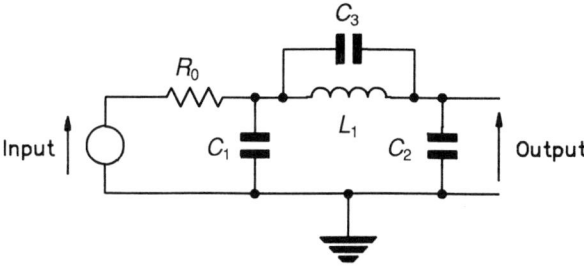

C_3

R_0

L_1

Input C_1 C_2 Output

5 The network shown in figure 9.16 is a 3rd-order low-pass singly-terminated ladder filter. Analysis of this network gives the following symbolic expression for the frequency response function:

Numerator:

$+1$

$+ L_1 C_3 (j\omega)^2$

Denominator:

$+1$

$+ R_0 C_1 (j\omega) + R_0 C_2 (j\omega)$

$+ L_1 C_2 (j\omega)^2 + L_1 C_3 (j\omega)^2$

$+ R_0 L_1 C_1 C_2 (j\omega)^3 + R_0 L_1 C_2 C_3 (j\omega)^3 + R_0 L_1 C_3 C_1 (j\omega)^3$

The filter is required to have an elliptic response with 1 dB passband ripple, 20 dB stopband attenuation and a cut-off frequency of 100 kHz:

$$H(j\omega) = \frac{1 + 1\cdot21 \times 10^{-12}(j\omega)^2}{1 + (2\cdot98 \times 10^{-6} j\omega) + (3\cdot69 \times 10^{-12}(j\omega)^2) + (6\cdot07 \times 10^{-18}(j\omega)^3)}$$

Taking R_0 to be 100 Ω, solve the simultaneous non-linear equations using the multidimensional Newton–Raphson algorithm to obtain the values of L_1, C_1, C_2 and C_3 that give the required response.

213

10 Tolerance Analysis

10.1 Introduction

In any mass-produced electronic system the component values are not exactly equal to their design values. Instead each component has a value which lies within a tolerance band around its nominal value. This is a consequence of the inevitable slight variability of the component manufacturing process and the effects of component ageing and temperature dependence. The behaviour of a system will therefore differ from the design behaviour, perhaps to the extent that it fails to meet its specification. In this case the manufacturing yield for the system will fall below 100%.

A yield of slightly less than 100% may be of little economic consequence. If a substantial proportion of systems fail to meet the specification, however, it will probably be necessary to reduce the tolerances on some of the component values. It is therefore important to be able to identify the most critical components, that is, the components whose values have the largest effect on the specification. Tighter component tolerances will, of course, lead to increased costs and it is quite possible that these will outweigh the savings brought about by the increased yield. Normally there will be a set of component tolerances that, while giving a yield of less than 100%, minimizes the overall manufacturing costs of those systems which meet the specification.

Reducing the component tolerances is not the only way to improve the yield, and two other approaches should be considered. In many cases the various parts of a specification are written without reference to their impact on the manufacturing costs. It may well happen that the effects of component tolerances are to violate only a small part of the specification, and that by relaxing this part of the specification an acceptable yield could be obtained without using more expensive components. The other approach is to consider whether an alternative network configuration would give a higher yield with the same component tolerances. A good example of this is in active filter design where the simplest method for realizing a required

Figure 10.1
A two-
dimensional
component
space

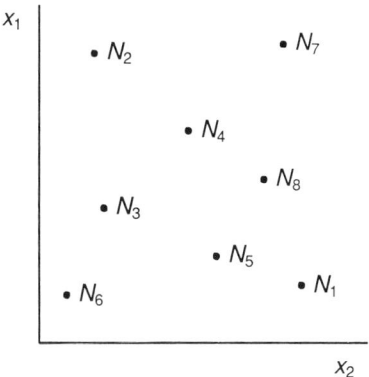

frequency response function is by a cascade of second-order stages. Unfortunately cascade filters are inordinately sensitive to component values and must be constructed from expensive close-tolerance components if a good yield is to be obtained. Passive equally-terminated ladder filters are relatively insensitive to component values but contain inductors and are impractical for use at frequencies below about 10 kHz. However, active filters derived by 'copying' passive ladder filters retain the same low sensitivity to component values as their passive prototypes, while eliminating the inductors. Such filters can replace cascade designs, giving an improved yield with the same component tolerances.

Suppose that an electronic network contains n components, each of which is specified by a single value. These component values define an n-dimensional space, the component space, in which a point represents some unique combination of component values. In other words, any network is represented by a single point in the component space. Consider a network containing only two components, of values x_1 and x_2. In this case the component space is a plane, as illustrated in figure 10.1. The points $N_1, N_2 \ldots N_8$ correspond to networks of the same configuration, but with different sets of component values.

In any real network the values of the components can assume only a limited range of values. Typically each component value x_i will lie within a tolerance band $\pm \Delta x_i$ centred on a nominal value x_i^0. This is illustrated for the case of two components in figure 10.2.

The volume of the component space bounded by the extreme values of the component values is known as the tolerance region (R_T) and all legitimate combinations of component values can be represented by points within the tolerance region.

Only some of the possible combinations of component values lead to an acceptable performance, that is a performance which meets all parts of the specification. Such component value combinations define a region of the component space known as the region of acceptability (R_A). A typical region of acceptability for a two-component network is shown in figure 10.3.

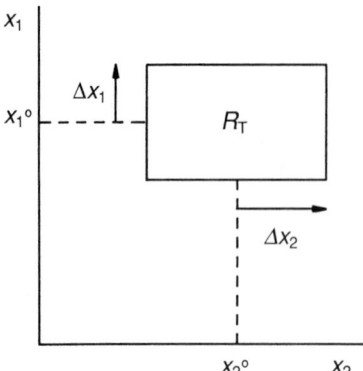

Figure 10.2
The tolerance region

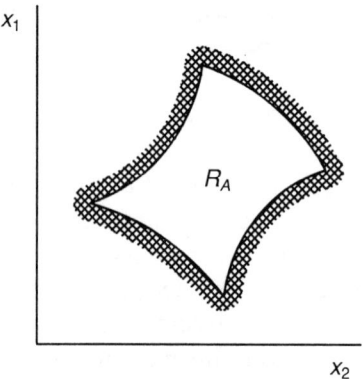

Figure 10.3
The region of acceptability

Unfortunately, except in very simple cases, there is no direct method for determining the region of acceptability from a given specification. Nevertheless it is a useful concept for discussing the effects of component tolerances. Figure 10.4 shows both the region of acceptability and the tolerance region for a two-component network.

In this example some of the tolerance region lies outside the region of acceptability and this implies that the manufacturing yield will be less than 100%. If the component value distributions are uniform, that is all

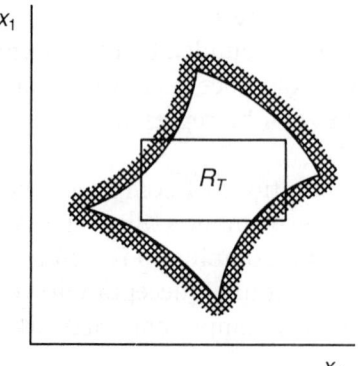

Figure 10.4
Manufacturing yield below 100%

216

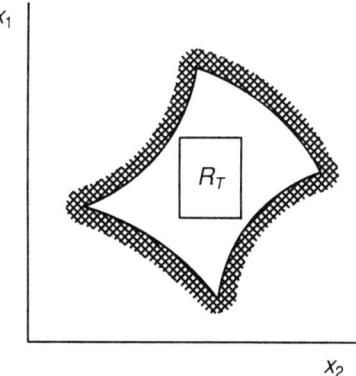

Figure 10.5
Manufacturing
yield of 100%

component values within the tolerance range are equally probable, then the failure rate is equal to the proportion of the volume of R_T that lies outside R_A. Normally, however, the distributions will be non-uniform and diagrams such as figure 10.4 must be considered to give only qualitative information concerning the yield.

By reducing the component tolerances the manufacturing yield can be improved up to 100%, as shown in figure 10.5. In this example it is sufficient to reduce the tolerance on x_2 only, in order to obtain a yield of 100%.

Methods for tolerance analysis fall into two groups. Worst-case analysis methods attempt to determine whether a network will meet its specification, even with the worst combination of component values. Worst-case methods do not give the yield if it is less than 100%. Statistical methods such as moments analysis and Monte-Carlo analysis predict, on the basis of known component value distributions, the manufacturing yield.

10.2 Worst-case Analysis

Worst-case tolerance analysis seeks to determine whether any combination of component values within the tolerance region leads to a performance that violates the specification. Only tolerance limits are used in worst-case analysis and no account is taken of the details of the component value distributions.

The n component values will be represented by a vector $\mathbf{x}$, where

$$\mathbf{x} = \begin{bmatrix} x_1 \\ x_2 \\ \vdots \\ x_n \end{bmatrix} \qquad (10.2.1)$$

In general a specification will cover several aspects of the performance, each of which can be measured by a performance function. Let f represent one such performance function; f will depend on all of the n component values:

$$f \equiv f(\mathbf{x}) \qquad (10.2.2)$$

The specification will usually place a lower limit l and an upper limit u on this performance function:

$$l \leqslant f(\mathbf{x}) \leqslant u \qquad (10.2.3)$$

A set of component values $\mathbf{x}^l$ within the tolerance region must exist, where

$$f(\mathbf{x}^l) \leqslant f(\mathbf{x}) \qquad (10.2.4)$$

for all $\mathbf{x}$ within the tolerance region. Also a set of components $\mathbf{x}^u$ within the tolerance region must exist, where

$$f(\mathbf{x}^u) \geqslant f(\mathbf{x}) \qquad (10.2.5)$$

for all $\mathbf{x}$ within the tolerance region. $\mathbf{x}^l$ and $\mathbf{x}^u$ are the worst-case component values for the performance function f. A manufacturing yield of 100% will therefore be obtained provided that

$$f(\mathbf{x}^l) \geqslant l \qquad (10.2.6)$$

and

$$f(\mathbf{x}^u) \leqslant u \qquad (10.2.7)$$

for all performance functions of the specification.

The problem lies, of course, in finding the worst-case component values $\mathbf{x}^l$ and $\mathbf{x}^u$. An obvious approach is to explore the extreme values of the components within the tolerance region. The performance function can be expanded as a first-order Taylor series around the nominal component valucs $\mathbf{x}^0$:

$$f(\mathbf{x}^0 + \delta\mathbf{x}) = f(\mathbf{x}^0) + \sum_{i=1}^{n} \frac{\partial f}{\partial x_i} \delta x_i \qquad (10.2.8)$$

where

$$\delta\mathbf{x} = \begin{bmatrix} \delta x_1 \\ \delta x_2 \\ \vdots \\ \delta x_n \end{bmatrix} \qquad (10.2.9)$$

Thus a small change $\delta\mathbf{x}$ in the component values leads to a change δf in the performance function given by

$$\delta f = \sum_{i=1}^{n} \frac{\partial f}{\partial x_i} \delta x_i \qquad (10.2.10)$$

If the component values are at the extremes of their tolerance bands, then

$$\delta x_i = \pm \Delta x_i \qquad (10.2.11)$$

Clearly the largest change Δf_{max} in the performance function will occur when the contributions from all of the components are of the same sign:

$$\Delta f_{max} = \sum_{i=1}^{n} \left| \frac{\partial f}{\partial x_i} \right| \Delta x_i \qquad (10.2.12)$$

The worst-case values of the performance function are then given by

$$f(\mathbf{x}^l) = f(\mathbf{x}^0) - \Delta f_{max} \qquad (10.2.13)$$

$$f(\mathbf{x}^u) = f(\mathbf{x}^0) + \Delta f_{max} \qquad (10.2.14)$$

In practice the partial derivatives of f with respect to the component values will have to be evaluated numerically:

$$\frac{\partial f}{\partial x_i} \approx \frac{f(x_1^0, \dots x_i^0 + \varepsilon_i, \dots x_n^0) - f(x_1^0, \dots x_i^0, \dots x_n^0)}{\varepsilon_i} \qquad (10.2.15)$$

It is convenient to choose the increments ε_i in equation (10.2.15) to be equal to the component tolerances Δx_i. Equation (10.2.12) then simplifies to

$$\Delta f_{max} \approx \sum_{i=1}^{n} |f(x_1^0, \dots x_i^0 + \Delta x_i, \dots x_n^0) - f(\mathbf{x}^0)| \qquad (10.2.16)$$

This expression is of course only approximate, relying as it does on a first-order Taylor expansion, and its use should be limited to situations where the component tolerances are relatively small.

A procedure *wcase1* which uses this method to calculate the worst-case values of a function f is shown below:

```
PROCEDURE wcase1;
VAR
    i, n: CARDINAL;
    x, dx: vector;
    f0, f1, df: REAL;
BEGIN
    init(x, dx, n);
    f0 := f(x, n); df := 0.0;
    WriteString('Nominal value = '); WriteReal(f0, 12); WriteLn;
    FOR i := 1 TO n DO
        x[i] := x[i]+dx[i];
        f1 := f(x, n);
        x[i] := x[i]-dx[i];
        df := df+ABS(f1-f0);
    END;
    WriteString('Highest value = '); WriteReal(f0+df, 12); WriteLn;
    WriteString('Lowest value =  '); WriteReal(f0-df, 12); WriteLn
END wcase1;
```

Figure 10.6
A Sallen–Key
active filter

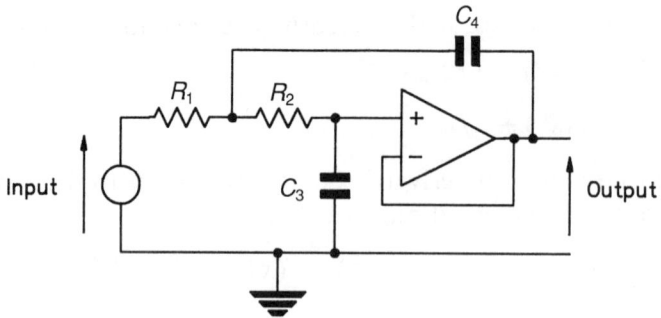

The procedure makes use of two other routines: *init* which sets up the nominal values x and the tolerances dx of the components, and f which evaluates the performance function.

To see how this method of worst-case analysis works in practice, consider the active filter network shown in figure 10.6. This network has a frequency-response function $H(j\omega)$ given by

$$H(j\omega) = \frac{1}{1 + C_3(R_1 + R_2)j\omega + R_1R_2C_3C_4(j\omega)^2}$$

With nominal component values of

$$
\begin{aligned}
R_1 &= 100 \ k\Omega & C_3 &= 1 \ nF \\
R_2 &= 100 \ k\Omega & C_4 &= 100 \ nF
\end{aligned}
$$

this network generates a low-pass response with a peak in the gain at the resonance frequency $\omega = 1000$ rad/s. Suppose that the worst-case values for the gain at $\omega = 1000$ rad/s are required, with resistor and capacitor tolerances of 5%. The procedure *wcase1* gives the following results:

$$
\begin{aligned}
\text{Nominal value} &= 1 \cdot 3979E + 001 \\
\text{Highest value} &= 1 \cdot 5837E + 001 \\
\text{Lowest value} &= 1 \cdot 2122E + 001
\end{aligned}
$$

Figure 10.7
Vertices of the
tolerance
region

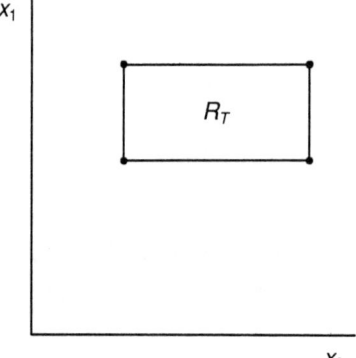

The maximum variation in the gain is $\pm 1 \cdot 9$ dB around the nominal value of $14 \cdot 0$ dB.

An alternative method of worst-case analysis, which can be used even when the tolerances are relatively large, is vertex analysis. In vertex analysis all of the combinations of extreme values of the components are investigated. These combinations of extreme values correspond to the vertices of the tolerance region, as illustrated for two components in figure 10.7.

The performance function f is evaluated at each of the vertices:

$$f(x_1 - \Delta x_1, \ x_2 - \Delta x_2, \ x_3 - \Delta x_3, \ \dots \ x_n - \Delta x_n)$$
$$f(x_1 + \Delta x_1, \ x_2 - \Delta x_2, \ x_3 - \Delta x_3, \ \dots \ x_n - \Delta x_n)$$
$$f(x_1 - \Delta x_1, \ x_2 + \Delta x_2, \ x_3 - \Delta x_3, \ \dots \ x_n - \Delta x_n)$$
$$f(x_1 + \Delta x_1, \ x_2 + \Delta x_2, \ x_3 - \Delta x_3, \ \dots \ x_n - \Delta x_n)$$
$$\vdots$$
$$f(x_1 + \Delta x_1, \ x_2 + \Delta x_2, \ x_3 + \Delta x_3, \ \dots \ x_n + \Delta x_n)$$

and a record is kept of the highest and lowest values of f.

A procedure *wcase2* which performs a full vertex analysis is shown below:

```
PROCEDURE wcase2;
VAR
    n: CARDINAL;
    x, dx: vector;
    fmin, fmax: REAL;

    PROCEDURE recursive(i: CARDINAL);
    VAR
        f0: REAL;
    BEGIN
        IF i <= n THEN
            x[i] := x[i]+dx[i];
            recursive(i+1);
            x[i] := x[i]-dx[i]-dx[i];
            recursive(i+1);
            x[i] := x[i]+dx[i]
        ELSE
            f0 := f(x, n);
            IF fmax < f0 THEN fmax := f0 END;
            IF fmin > f0 THEN fmin := f0 END;
        END
    END recursive;
```

```
BEGIN
    init(x, dx, n);
    fmax := f(x, n); fmin := fmax;
    WriteString('Nominal value = '); WriteReal(fmax, 12); WriteLn;
    recursive(1);
    WriteString('Highest value = '); WriteReal(fmax, 12); WriteLn;
    WriteString('Lowest value =  '); WriteReal(fmin, 12); WriteLn
END wcase2;
```

___

The local procedure *recursive* is called at n different levels, corresponding to the n components; at each level i, it generates the extreme values $x_i \pm \Delta x_i$ of the ith component. When applied to the active filter network of figure 10.6, with resistor and capacitor tolerances of 5%, the procedure *wcase2* gave the following results:

Nominal value $= 1 \cdot 3979E + 001$
Highest value $\; = 1 \cdot 4422E + 001$
Lowest value $\; = 1 \cdot 0220E + 001$

There is a significant difference between these results, and those obtained by using partial derivatives. In particular the effects of component tolerances are seen to be asymmetric with a positive variation in gain of $0 \cdot 4$ dB, but a negative variation of $-3 \cdot 8$ dB.

Vertex analysis is limited in its application by the amount of computation that is necessary to evaluate f at all of the vertices. Each component has two extreme values and a network of n components therefore has a tolerance region with 2^n vertices. This is acceptable in the case of a network of 10 components where 1024 evaluations of a performance function might take a few seconds on a personal computer. On the other hand, vertex analysis of a network of 20 components involves more than a million evaluations of the performance function and this would probably take several hours on a personal computer.

There is another drawback to vertex analysis. Intuitively one might expect that the worst performance would occur at the extreme component values, but this is not necessarily the case. The region of acceptability may be quite complex, and a situation similar to that illustrated in figure 10.8 can arise. Here the vertices of the tolerance region all lie within the region of acceptability. Nevertheless, part of the tolerance region extends outside the region of acceptability and the manufacturing yield would be less than 100%.

Worst-case tolerance analysis methods tend to be pessimistic. A network may fail to meet its specification if all the components have values close to their extremes, but the likelihood of this happening is probably very small.

Figure 10.8
Failure of
vertex analysis

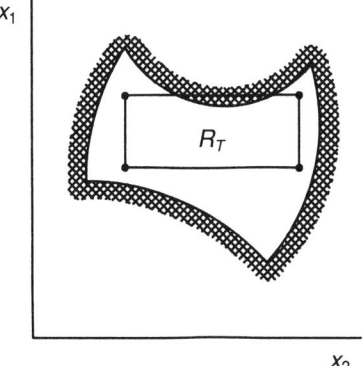

Worst-case analysis is not, however, concerned with probabilities and simply indicates whether a specification is violated. This might lead to tighter tolerances being imposed on the component values, even though the manufacturing yield with the original tolerances was, say, 99·99%. As a result the overall manufacturing costs would be increased unnecessarily.

10.3 Component Value Distributions

All statistical tolerance analysis methods demand a knowledge of the component value distributions. Although manufacturers give nominal values and tolerances for their components, they rarely provide details of the component value distributions.

Suppose that one thousand resistors, of nominal value 100 kΩ and tolerance 5%, were to be measured and the results plotted as a histogram. The result might be similar to that shown in figure 10.9. Because of the limited number of components tested each cell of the histogram exhibits a

Figure 10.9
Component
value histogram

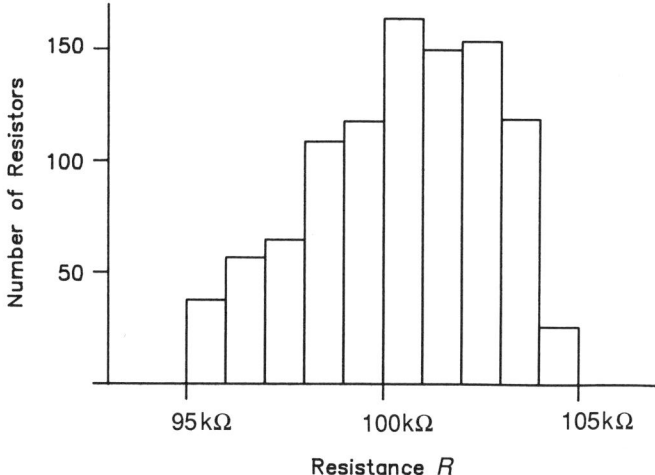

223

Figure 10.10
Probability
density
function

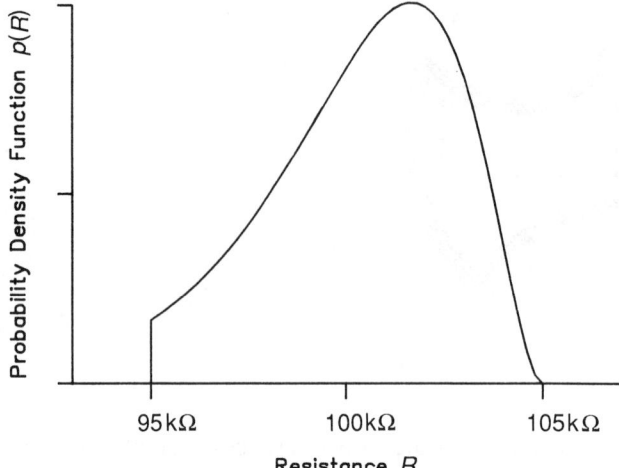

statistical variation and the envelope of the histogram is not a smooth curve. Increasing the number of components to be measured leads to a reduction in the statistical effects, and allows the tolerance range to be split into a larger number of intervals. Ultimately, as the number of components becomes infinite, the histogram turns into a continuous function as shown in figure 10.10.

This function is known as the probability density function $p(x)$, and $p(x)\,\mathrm{d}x$ is the probability that a component value lies between x and $x + \mathrm{d}x$. Clearly all component values lie between $-\infty$ and $+\infty$ so that the probability of any component value falling within this range, which is equal to the total area under the probability density function, is simply unity:

$$\int_{-\infty}^{\infty} p(x)\,\mathrm{d}x = 1 \tag{10.3.1}$$

In practice component value distributions can have many different forms. During the manufacturing process several different factors can affect the final component values, and these factors will usually combine to give an approximately Gaussian (or normal) distribution:

$$p(x) = \frac{1}{\sigma\sqrt{(2\pi)}}\,\exp\left(-\frac{(x-\mu)^2}{2\sigma^2}\right) \tag{10.3.2}$$

where μ is the mean and σ the standard deviation of the distribution. Figure 10.11 shows a typical Gaussian distribution of component values with a tolerance of $\pm 10\%$.

It is often uneconomic to produce components to a particular tolerance specification without selection. A manufacturer making 5% resistors might therefore take components with a distribution similar to that shown in figure 10.11 and discard the small proportion whose values fall outside the 5% tolerance range. The result would be a truncated Gaussian distribution as shown in figure 10.12.

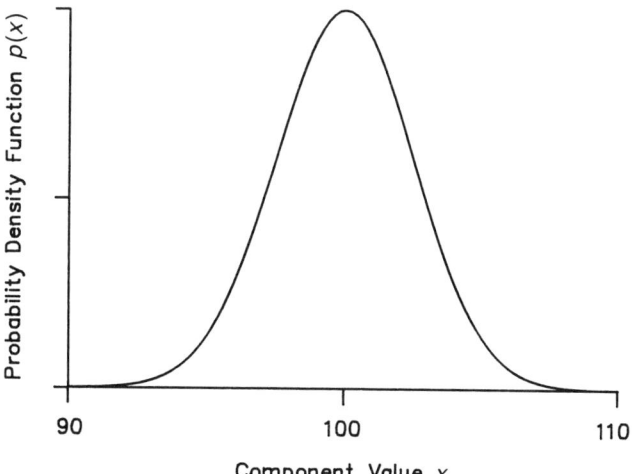

Figure 10.11
A Gaussian
distribution

Figure 10.12
A truncated
Gaussian
distribution

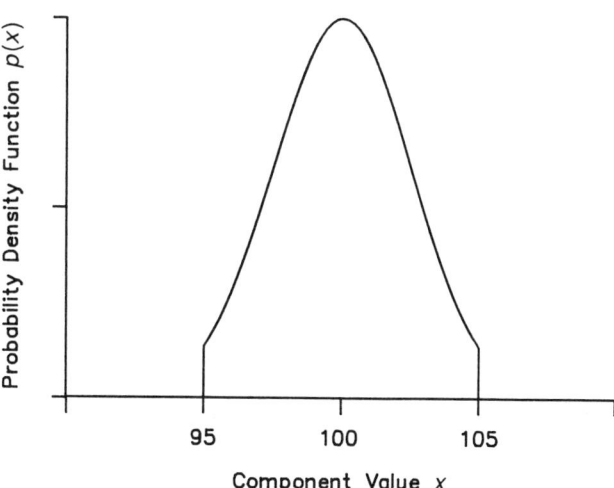

If components are required with a very close tolerance, then these will almost certainly be selected from devices with a considerably greater manufacturing tolerance. For example, 1% resistors might be selected from the distribution shown in figure 10.11 and sold at a premium. Since they are selected from the central part of a Gaussian distribution, the values of these resistors will be approximately uniformly distributed as shown in figure 10.13.

Finally, the components remaining after selection of the close-tolerance devices will have a bimodal distribution as shown in figure 10.14.

The mean μ is one measure of the centre of a distribution (the others are the median and the mode) and is defined to be

$$\mu = \int_{-\infty}^{\infty} x p(x) \, \mathrm{d}x \tag{10.3.3}$$

225

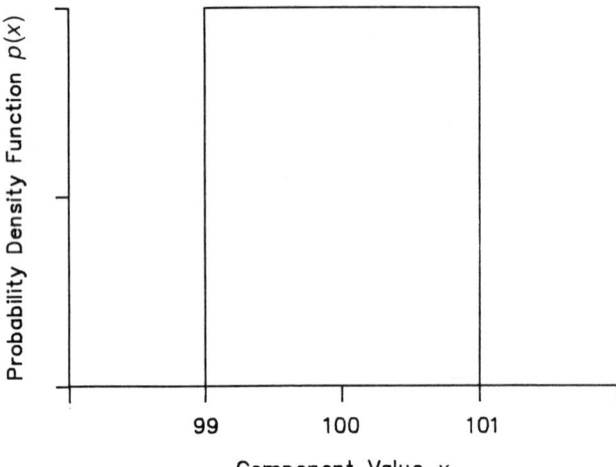

Figure 10.13
A uniform
distribution

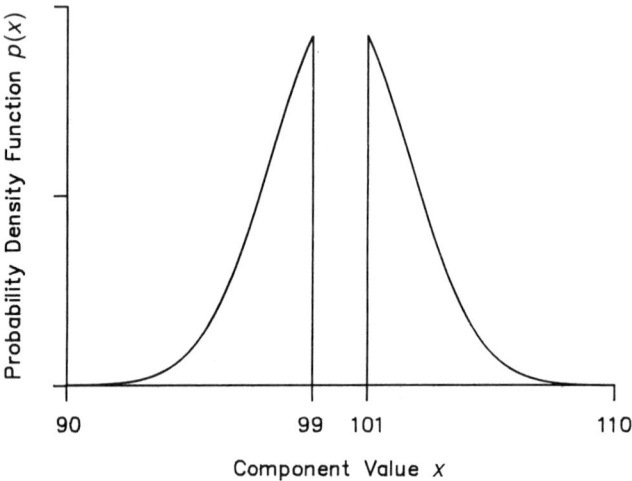

Figure 10.14
A bimodal
distribution

A distribution can be characterized by its moments. The rth moment μ_r about the mean of the distribution is given by

$$\mu_r = \int_{-\infty}^{\infty} (x - \mu)^r p(x) \, \mathrm{d}x \qquad (10.3.4)$$

Because of its importance in measuring the width of a distribution a special name, the variance, is given to the second moment μ_2 of a distribution about its mean. An alternative measure of the width is the standard deviation σ, which is equal to the square root of the variance:

$$\sigma^2 = \mu_2 = \int_{-\infty}^{\infty} (x - \mu)^2 p(x) \, \mathrm{d}x \qquad (10.3.5)$$

In the discussion of tolerance methods which follows it will be assumed that the component values in a network are uncorrelated. In other words each component value is determined by the probability density function for that type of component, and is independent of all other component values. This is normally a valid assumption for networks constructed from discrete components, but not for integrated circuits. An integrated circuit is fabricated, together with hundreds or thousands of similar devices, on a wafer of semiconductor material. Component values may vary considerably over a wafer, or between different wafers, so that the overall tolerance might be, for example, 20%. Within a particular integrated circuit, however, the variation is likely to be much less, with values matching to, say, 2%. The component values are therefore highly correlated.

10.4 Method of Moments

It will be assumed that some aspect of the performance of a network is measured by a performance function f which depends on all of the component values $x_1 \ldots x_n$. The method of moments is based on the fact that a probability density function is completely characterized by its mean and moments. If the dependence of f on the component values is known then it should be possible to calculate the moments of the distribution of f from the moments of the component value distributions. The distribution of f is thus determined, and from both this and the specification can be deduced the manufacturing yield. In practice a number of assumptions are made in calculating the moments and the method of moments gives only approximate results. It is, however, much more efficient than alternative statistical techniques.

Each of the components has a random value within its tolerance range and influences the performance function f. A theorem of statistics, the Central Limit Theorem, states that the probability density function of a sum of random variables will tend to be Gaussian, provided that the variances of the random variables are similar. Surprisingly this result is independent of the probability density functions of the variables. It is to be expected, therefore, that the probability density function of f will be approximately Gaussian. This fact greatly simplifies tolerance analysis by the method of moments because a Gaussian probability density function is completely determined by its mean and variance.

Let the vector $\mathbf{x}^0$ represent the mean values of the components. Since component distributions are usually symmetrical it will be assumed that $\mathbf{x}^0$ is equal to the nominal component values. The performance function can be expanded around the point $\mathbf{x}^0$ as a first-order Taylor series:

$$f(\mathbf{x}^0 + \delta\mathbf{x}) = f(\mathbf{x}^0) + \sum_{i=1}^{n} \frac{\partial f}{\partial x_i} \delta x_i \qquad (10.4.1)$$

Now suppose that f is evaluated for a large number m of sets of random component values. The mean μ of f is given by

$$\mu = \frac{1}{m} \sum_{j=1}^{m} f(\mathbf{x}^j) \tag{10.4.2}$$

where $\mathbf{x}^j$ is the jth set of random values. Substituting the Taylor expansion into this expression gives

$$\mu = \frac{1}{m} \sum_{j=1}^{m} \left\{ f(\mathbf{x}^0) + \sum_{i=1}^{n} \frac{\partial f}{\partial x_i} \delta x_i^j \right\} \tag{10.4.3}$$

where δx_i^j represents the jth deviation of the ith component from its mean value x_i^0. This expression can be rearranged to give

$$\mu = f(\mathbf{x}^0) + \sum_{i=1}^{n} \frac{\partial f}{\partial x_i} \left\{ \frac{1}{m} \sum_{j=1}^{m} \delta x_i^j \right\}$$

$$\approx f(\mathbf{x}^0) \quad \text{for large } m \tag{10.4.4}$$

In other words the mean value of f is simply the value of f evaluated for the mean component values. A similar procedure can be used to derive the variance σ_f^2 of the performance function.

$$\sigma_f^2 = \frac{1}{m} \sum_{j=1}^{m} \{f(\mathbf{x}^j) - \mu\}^2 \tag{10.4.5}$$

$$= \frac{1}{m} \sum_{j=1}^{m} \left\{ \sum_{i=1}^{n} \frac{\partial f}{\partial x_i} \delta x_i^j \right\}^2 \tag{10.4.6}$$

It will now be assumed that the component values are uncorrelated, that is, if $i \neq k$ then

$$\frac{1}{m} \sum_{j=1}^{m} \delta x_i^j \delta x_k^j \to 0 \qquad \text{as } m \to \infty \tag{10.4.7}$$

This leads to a simple expression for the variance of f:

$$\sigma_f^2 \approx \sum_{i=1}^{n} \left(\frac{\partial f}{\partial x_i} \right)^2 \frac{1}{m} \sum_{j=1}^{m} (\delta x_i^j)^2 \qquad \text{for large } m$$

$$= \sum_{i=1}^{n} \left(\frac{\partial f}{\partial x_i} \right)^2 \sigma_i^2 \tag{10.4.8}$$

where σ_i^2 is the variance of the ith component value.

In practice the partial derivatives of f must be evaluated numerically:

$$\frac{\partial f}{\partial x_i} \approx \frac{f(x_1^0, \ldots x_i^0 + \varepsilon_i, \ldots x_n^0) - f(x_1^0, \ldots x_i^0, \ldots x_n^0)}{\varepsilon_i} \tag{10.4.9}$$

It is convenient to choose the increments ε_i in equation (10.4.9) to be equal to the standard deviations σ_i of the component values. Equation (10.4.8) then becomes:

$$\sigma_f^2 \approx \sum_{i=1}^{n} \{f(x_1^0 \ldots x_i^0 + \sigma_i \ldots x_n^0) - f(\mathbf{x}^0)\}^2 \qquad (10.4.10)$$

The probability density function $p(f)$ of the performance function f is approximately Gaussian:

$$p(f) = \frac{1}{\sigma_f \sqrt{(2\pi)}} \exp\left(-\frac{(f-\mu)^2}{2\sigma_f^2}\right) \qquad (10.4.11)$$

where μ are σ are obtained using equations (10.4.4) and (10.4.10). From this distribution the probability of f falling outside the specification can be determined. Suppose that l and u represent the lower and upper specification limits on f:

$$l \leqslant f \leqslant u \qquad (10.4.12)$$

The probability that the network will fail to meet its specification is proportional to the area of $p(f)$ which is outside these limits. This is illustrated in figure 10.15. The probability of failure p_l through f being below the lower limit l is given by

$$p_l = \int_{-\infty}^{l} p(f) \, \mathrm{d}f \qquad (10.4.13)$$

and the probability of failure p_u through f being above the upper limit u is given by

$$p_u = \int_{u}^{\infty} p(f) \, \mathrm{d}f \qquad (10.4.14)$$

Unfortunately a Gaussian function cannot be integrated algebraically and in order to evaluate p_l and p_u it is necessary to resort to numerical methods.

Figure 10.15
Probability of
failure

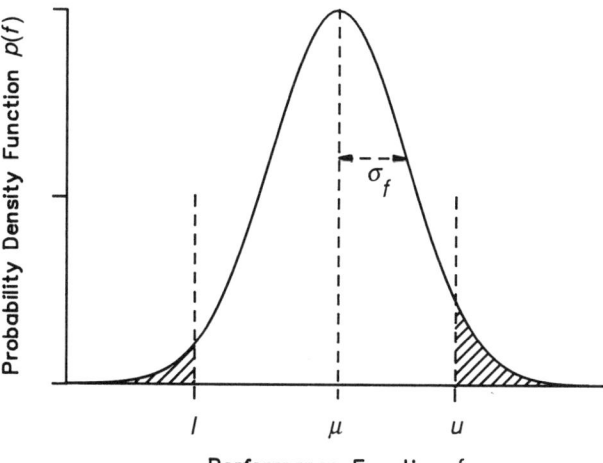

229

It is convenient at this point to introduce the complementary error function erfc(x), where

$$\text{erfc}(x) = \frac{2}{\sqrt{\pi}} \int_x^\infty \exp(-y^2)\, \mathrm{d}y \qquad\qquad (10.4.15)$$

The numerical integration procedure *adaptive* described in chapter 7 can be used to evaluate this function:

```
PROCEDURE g(y: REAL): REAL;
BEGIN
    RETURN exp(-y*y)
END g;


PROCEDURE erfc(x: REAL): REAL;
BEGIN
    RETURN 2.0*adaptive(g, x, 10.0, 1.0E-8)/sqrt(pi)
END erfc;
```

Finally the failure probabilities p_l and p_u can be written in terms of the complementary error function:

$$p_l = \tfrac{1}{2}\,\text{erfc}\,\frac{\mu - l}{\sigma_f\sqrt{2}} \qquad\qquad (10.4.16)$$

$$p_u = \tfrac{1}{2}\,\text{erfc}\,\frac{u - \mu}{\sigma_f\sqrt{2}} \qquad\qquad (10.4.17)$$

The procedure *moments* shown below performs a method-of-moments analysis and prints the failure rates. It assumes that the component value distributions are uniform, extending from $x_i - \Delta x_i$ to $x_i + \Delta x_i$. Such a distribution has a standard deviation $\sigma_i = \Delta x_i / \sqrt{3}$.

```
PROCEDURE moments(1, u: REAL);
VAR
    i, n: CARDINAL;
    x, dx: vector;
    f0, f1, v, si, pl, pu, q: REAL;
BEGIN
    init(x, dx, n);
    f0 := f(x, n); v := 0.0;
    WriteString('Nominal value = '); WriteReal(f0, 12); WriteLn;
```

```
FOR i := 1 TO n DO
    q := dx[i]/sqrt(3.0);
    x[i] := x[i]+q;
    f1 := f(x, n);
    x[i] := x[i]-q;
    v := v+(f1-f0)*(f1-f0);
END;
si := sqrt(v);
WriteString('Sigma =           '); WriteReal(si, 12); WriteLn;
pl := erfc((f0-1)/(si*sqrt(2.0)))/2.0;
pu := erfc((u-f0)/(si*sqrt(2.0)))/2.0;
WriteString('Failure rate low =   ');
WriteFix(100.0*pl, 2, 4); WriteString(' %'); WriteLn;
WriteString('Failure rate high = ');
WriteFix(100.0*pu, 2, 4); WriteString(' %'); WriteLn;
WriteString('Total failure rate = ');
WriteFix(100.0*(pl+pu), 2, 4); WriteString(' %'); WriteLn
END moments;
```

This procedure was used to perform a tolerance analysis on the active filter network of figure 10.6 with resistor and capacitor tolerances of 5%. Upper and lower specification limits on the gain were set at $13 \cdot 0$ dB and $15 \cdot 0$ dB and the following results were obtained:

Nominal value $= 1 \cdot 3979\mathrm{E} + 001$
Sigma $= 4 \cdot 5564\mathrm{E} - 001$
Failure rate low $= 1 \cdot 5798\%$
Failure rate high $= 1 \cdot 2548\%$
Total failure rate $= 2 \cdot 8346\%$

In fact these results underestimate the true failure rate. There are two reasons for this. The small number of components means that the distribution of f is a poor approximation to Gaussian. Also the fairly high component tolerances result in the first-order Taylor expansion used to derive expressions for μ and σ_f being inadequate.

So far it has been assumed that only one performance function is of interest. In practice, however, there may be several functions which must simultaneously meet their specification limits. Unfortunately there is no simple way of combining the failure probabilities for the different functions because the functions are likely to be highly correlated. The method of moments is therefore most suitable for rapid estimation of the failure probability for a single performance function.

10.5 Monte-Carlo Analysis

Monte-Carlo analysis is a statistical technique that estimates the manufacturing yield by simulating the process of manufacturing an electronic system. It is the most accurate and reliable of the tolerance analysis methods because it does not involve approximations. In particular there is no assumption of linearity between the component values and the performance functions so that the effects of large component tolerances are correctly determined. Since it is a statistical method it can only give an estimate of the yield, although it is possible to calculate the degree of confidence that can be placed in the results.

During the manufacture of an electronic system each component is selected at random from a batch of similar components of suitable nominal value. Monte-Carlo analysis simulates the random selection process on a computer by the use of a random number generator with an appropriate probability density function. Once a complete set of component values has been selected the various performance factors making up the specification can be evaluated. If any one of these functions falls outside its specification limits then the system is regarded as having failed the simulated manufacturing process. This complete process must be repeated a large number of times and the proportion of failures can be used to estimate the actual manufacturing yield. The operation of the Monte-Carlo method is illustrated in figure 10.16.

There are twenty sets of component values x_1 and x_2, chosen randomly with uniform probability density functions, in the tolerance region. Of these, all except five are also within the region of acceptability. The failure rate is therefore 25% and the simulated manufacturing yield, which is the complement of the failure rate, is 75%.

Obviously a central problem in Monte-Carlo analysis is the generation of random numbers with the required probability density functions. Small departures from the ideal in the random number generator, for example correlations between successive random numbers, can lead to serious errors

Figure 10.16
The Monte-Carlo method

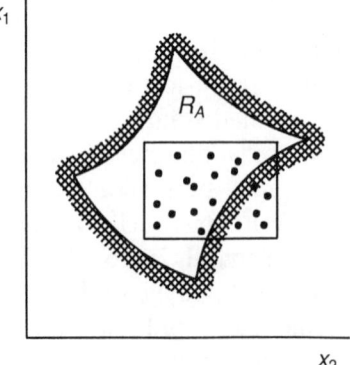

in the results of the Monte-Carlo analysis. Random number generation is dealt with in detail in sections 10.6 and 10.7.

The procedure *montecarlo* given below estimates the failure rate for a system specified by a single performance function f which is required to lie between l and u. The parameter N determines the number of sets of component values to be analyzed.

```
PROCEDURE montecarlo(l, u: REAL; N: CARDINAL);
VAR
    i, j, n, flo, fhi: CARDINAL;
    x, dx, y: vector;
    f1: REAL;
BEGIN
    init(x, dx, n);
    flo := 0; fhi := 0;
    FOR i := 1 TO N DO
        FOR j := 1 TO n DO y[j] := uniform(x[j], dx[j]) END;
        f1 := f(y, n);
        IF f1 < l THEN INC(flo) END;
        IF f1 > u THEN INC(fhi) END
    END;
    WriteString('Number of trials = '); WriteCard(N, 6); WriteLn;
    WriteString('Failed low =      '); WriteCard(flo, 6); WriteLn;
    WriteString('Failed high =     '); WriteCard(fhi, 6); WriteLn;
    WriteString('Total failure rate = ');
    WriteFix(100.0*FLOAT(fhi+flo)/FLOAT(N), 2, 4);
    WriteString(' %'); WriteLn;
END montecarlo;
```

This procedure makes use of a routine *uniform* which returns a random number with a uniform distribution over the range $x - dx$ to $x + dx$.

Monte-Carlo analysis of the active filter network shown in figure 10.6, with 5% tolerance components and specification limits of $13 \cdot 0$ dB and $15 \cdot 0$ dB, gave the following results:

Number of trials = 1000
Failed low = 103
Failed high = 0
Total failure rate = $10 \cdot 3000\%$

This analysis took 80 sec on a standard IBM personal computer.

The true failure probability p_∞ is defined to be the proportion of random component value combinations that would fail to meet the specification in an infinite number of trials. How then does the failure rate $p = 10 \cdot 3\%$ obtained for 1000 trials relate to p_∞? Further Monte-Carlo analyses, each involving 1000 trials, gave failure rates of $p = 9 \cdot 4\%$, $p = 10 \cdot 6\%$, $p = 8 \cdot 8\%$ and $p = 9 \cdot 9\%$. The average of these results, $p = 9 \cdot 8\%$, is likely to be a better approximation to p_∞ than the individual results, but is derived from a total of 5000 trials. To see what significance can be attached to these failure rates, the 1000-trial Monte-Carlo analysis was repeated one hundred times. The mean of the estimated failure rates p was found to be $\mu_p = 9 \cdot 80\%$ and the standard deviation $\sigma_p = 0 \cdot 96\%$. Figure 10.17 shows the failure rates plotted as a histogram.

The shape of the histogram suggests that the distribution $\psi(p)$ of estimated failure rates p is Gaussian. This is to be expected because each of the estimated failure rates is the sum of contributions q_i from a large number N of component value combinations:

$$p = \frac{1}{N} \sum_{i=1}^{N} q_i \tag{10.5.1}$$

where q_i is 0 if the ith trial meets the specification and 1 if it fails. Now the Central Limit Theorem states that the distribution of a sum of a large number of random variables will tend to be Gaussian. If the 1000-trial Monte-Carlo analysis could be performed an infinite number of times, therefore, a Gaussian probability density function similar to that shown in figure 10.18 would be obtained.

The mean value of p, that is the centre of the distribution, is the mean of q_i averaged over the infinite number of trials. This by definition is the true failure rate p_∞. What then is the probability that a single N-trial Monte-Carlo analysis will underestimate the true failure rate p_∞ by an

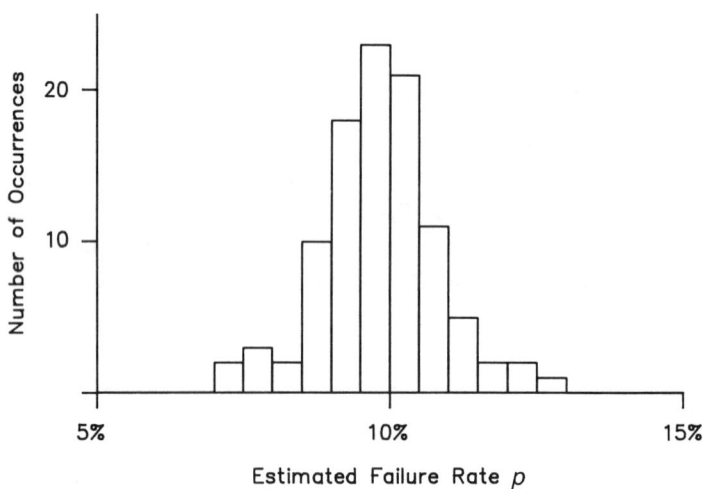

Figure 10.17
Histogram of estimated failure rates

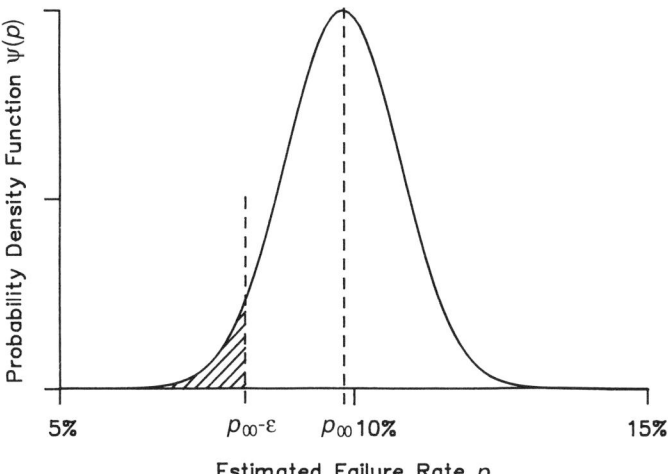

Figure 10.18
Distribution of estimated failure rates

amount ε? From figure 10.18 it is clear that the probability of underestimation P_ε is equal to the area under the Gaussian function below $p_\infty - \varepsilon$:

$$P_\varepsilon = \int_{-\infty}^{p_\infty - \varepsilon} \psi(p) \, \mathrm{d}p \qquad (10.5.2)$$

$$= \tfrac{1}{2} \operatorname{erfc} \frac{\varepsilon}{\sigma_p \sqrt{2}} \qquad (10.5.3)$$

Suppose that a Monte-Carlo analysis produces an estimated failure rate p. Then it is possible to say that

$$p_\infty < p + \varepsilon \qquad (10.5.4)$$

with a confidence level $1 - P_\varepsilon$. By a similar argument it is also possible to say that

$$p_\infty > p - \varepsilon \qquad (10.5.5)$$

with the same confidence level $1 - P_\varepsilon$. Usually, however, the lower limit on p_∞ is less important than the upper limit.

If $\varepsilon = 2\sigma_p$ then the probability of underestimation given by equation (10.5.3) is $P_\varepsilon = 2 \cdot 3\%$. Thus

$$p_\infty < p + 2\sigma_p \qquad (10.5.6)$$

at the 97% confidence level. If $\varepsilon = 3\sigma_p$ then the probability of underestimation is $P_\varepsilon = 0 \cdot 13\%$ and

$$p_\infty < p + 3\sigma_p \qquad (10.5.7)$$

at the 99·8% confidence level.

The remaining problem is to determine the standard deviation σ_p of p. One approach is to repeat the N-trial Monte-Carlo analysis a number of

235

times and calculate the standard deviation of the results. When this was done for the 1000-trial analysis of the active filter a value $\sigma_p = 0 \cdot 96\%$ was obtained. However it is possible to estimate σ_p much more efficiently from a single N-trial Monte-Carlo analysis. It is easy to show that for a failure rate p of the form of the equation (10.5.1) the variance is given by

$$\sigma_p^2 = \sigma_q^2 / N \tag{10.5.8}$$

where σ_q^2 is the variance of the individual trials. Now q has a value 1 with probability p_∞, and a value 0 with probability $(1 - p_\infty)$, so that the variance of q is given by

$$\sigma_q^2 = p_\infty (1 - p_\infty)^2 + (1 - p_\infty)(0 - p_\infty)^2$$
$$= p_\infty (1 - p_\infty) \tag{10.5.9}$$

Finally σ_p^2 can be obtained in terms of p_∞ by combining equations (10.5.8) and (10.5.9):

$$\sigma_p^2 = p_\infty (1 - p_\infty) / N \tag{10.5.10}$$

Of course p_∞ itself is not known, but the estimate p from the Monte-Carlo analysis should be a reasonable approximation to p_∞ and can replace it in equation (10.5.10):

$$\sigma_p^2 \approx p(1 - p) / N \tag{10.5.11}$$

The result of the 1000-trial Monte-Carlo analysis of the active filter was an estimated failure rate of $10 \cdot 3\%$. Using equation 10.5.11 gives the standard deviation $\sigma_p = 0 \cdot 96\%$. The true failure rate can therefore be taken to be less than $12 \cdot 2\%$ at the 97% confidence level, and less than $13 \cdot 2\%$ at the $99 \cdot 8\%$ confidence level.

10.6 Random Number Generation

Random numbers are usually generated initially with a uniform distribution (that is with a constant probability density function $p(x)$) over the range 0 to 1; any other distribution can then be obtained by applying a suitable transformation. When a random number is generated as part of a uniform distribution, any value within the range is equally likely, and the new random number must be entirely independent of previously generated numbers. This effectively rules out the use of normal computers which are (or at least are designed to be) deterministic machines to generate random numbers. It is, of course, possible to construct true random number generators which exploit certain random physical effects (such as radioactive decay) but these are not widely available.

Fortunately, for the purposes of Monte-Carlo analysis it is only necessary to generate numbers which have properties similar to true random numbers. Such numbers are known as pseudo-random numbers and can be generated by an algorithm running on a deterministic machine. If, according to various appropriate tests, a sequence of numbers does not differ significantly from true random numbers then it can be considered to be satisfactory for the purposes of Monte-Carlo analysis.

The difficulty lies in deciding what tests to apply. Obviously the mean and variance must correspond with those of a uniform distribution (mean = 1/2, variance = 1/12) but these criteria could be satisfied by a distribution consisting entirely of pairs of numbers equal to $1/2 \pm 1/\sqrt{12}$. Higher-order moments could be evaluated to give a more sensitive test of the uniformity of the distribution but such results are difficult to interpret.

An alternative way of testing for uniformity is to divide the range 0 to 1 into a number of subranges, for example 0 to $0 \cdot 1$, $0 \cdot 1$ to $0 \cdot 2$, ... $0 \cdot 9$ to $1 \cdot 0$. Each pseudo-random number falls into one of these subranges. After a sufficient number N of random values have been accumulated the contents p_i of each of the subranges can be compared with their ideal values $q_i = N/n$ where n is the number of subranges. To establish whether the results differ significantly from true random numbers the chi-square test can be applied. χ^2 is evaluated where

$$\chi^2 = \sum_{i=1}^{n} \frac{(p_i - q_i)^2}{q_i} \qquad (10.6.1)$$

Published tables give the value of χ^2 for different levels of significance, with different numbers of degrees of freedom. (In this case the number of degrees of freedom is $n - 1$).

It is not, of course, sufficient simply to test for uniformity; the series of numbers $0 \cdot 00$, $0 \cdot 01$, $0 \cdot 02$, ... $0 \cdot 99$ has perfect uniformity but is far from being random. Nor is it sufficient to test for serial correlation between successive numbers because complex correlations involving several numbers can occur. A useful empirical test for measuring correlation is the so-called poker test in which groups of five consecutive numbers are regarded as hands in the card game of poker. Each number is placed in one of the ten subranges 0 to $0 \cdot 1$, $0 \cdot 1$ to $0 \cdot 2$, ... $0 \cdot 9$ to $1 \cdot 0$, and the groups of five numbers are categorized according to whether there is a pair, three, four or five numbers in any subrange, or whether all five fall into different subranges. The test is repeated for a large number of groups (at least 100 000 groups should be used to obtain good statistics) and the results compared with those for true random numbers. The chi-square test can then be used to establish whether the results differ significantly from true random numbers.

All pseudo-random number generators are cyclic, that is after a certain number of values the sequence repeats. The length of the subsequence, which by repetition forms the entire sequence, is known as the period. Obviously the period should be greater than the largest number of random values that are likely to be used in a Monte-Carlo analysis. A period of 10^6

should be considered to be the minimum for a general-purpose pseudo-random number generator.

The earliest pseudo-random number generators were based on the so-called mid-square principle in which a new random value is produced by squaring the previous value and selecting the middle digits of the result. Most modern generators operate on either the congruential or the feedback shift register principles. In both cases random integers over a certain range are generated and these are then scaled to give real results in the interval 0 to 1. Provided that the range of the integers is large enough then the real results form a good approximation to a continuous distribution. Feedback shift register methods involve both arithmetic and bit manipulations; they are most suitable for implementation at machine code level and will not be discussed further.

Congruential pseudo-random number generators employ only arithmetic operations and are readily programmed in high-level languages. The simple formula given below is used to generate a new integer x_{n+1} from the previous one x_n.

$$x_{n+1} = (ax_n + c) \bmod m \tag{10.6.2}$$

Integers produced by this formula lie in the range 0 to $m - 1$; the period of a generator based on this formula cannot therefore exceed m. Each integer is divided by the modulus m to give a real number in the range 0 to 1. It is usual (although not necessary) to make the modulus m equal to a power of 2. The multiplier a and the increment c are to some extent arbitrary, although they should both be less than, and relatively prime to, the modulus m. Congruential pseudo-random number generators fall into two classes depending on the value of c.

In a multiplicative congruential generator the increment c is zero. If m is a power of 2, then the maximum period of such a generator is $m/4$ and is obtained when the multiplier a is given by

$$a = 8k + 3 \quad \text{or} \quad a = 8k + 5$$

where k is a positive integer. It is essential that the initial value x_0 is odd, otherwise the maximum period will not be achieved. In particular, if the value of x_0 is zero, then all succeeding values x_i will also be zero.

Mixed congruential generators have a non-zero increment c. If m is a power of 2, c is odd, and the multiplier a is given by

$$a = 4k + 1$$

then the period is equal to m. During a complete period therefore each of the values from 0 to $m - 1$ will appear once and the pseudo-random sequence has perfect uniformity. Obviously, in this type of generator any starting value in the range 0 to $m - 1$ can be used. Normally the mixed congruential method is to be preferred to the multiplicative congruential method on account of its longer period and insensitivity to starting values.

Table 10.1

i	x_i	x_i/m
1	0	0·0
2	7	0·21875
3	10	0·3125
4	25	0·78125
5	4	0·125
6	27	0·84375
7	14	0·4375
8	13	0·40625
9	8	0·25
10	15	0·46875
11	18	0·5625
12	1	0·03125
13	12	0·375
14	3	0·09375
15	22	0·6875
16	21	0·65625
17	16	0·5
18	23	0·71875
19	26	0·8125
20	9	0·28125
21	20	0·625
22	11	0·34375
23	30	0·9375
24	29	0·90625
25	24	0·75
26	31	0·96875
27	2	0·0625
28	17	0·53125
29	28	0·875
30	19	0·59375
31	6	0·1875
32	5	0·15625
33	0	0·0

The sequence generated by a mixed congruential generator with $m = 32$, $a = 5$ and $c = 7$ is shown in Table 10.1.

Although the modulus m is far smaller than would be employed in any practical generator, there is no apparent pattern in the sequence of integers, except that they alternate between being even and odd.

A considerable amount of work has been done on selecting the best values of a and c. Provided that a is small compared with $\sqrt{m}$ then the serial correlation ρ between successive numbers is given approximately by

$$\rho = \frac{1}{a} - \frac{6c}{am}\left(1 - \frac{c}{m}\right) \tag{10.6.3}$$

This equation shows that the largest value of a that is small compared with $\sqrt{m}$ should be used. It also implies that the best value of the increment c

239

(giving a zero correlation) is independent of a and is given by

$$c = \frac{m}{2} \left(1 - \frac{1}{\sqrt{3}} \right) \qquad (10.6.4)$$

A congruential pseudo-random number generator meeting the above criteria, and with a period large enough for any Monte-Carlo analysis likely to be performed on a personal computer, can be based on the following parameters

$$m = 4294967296 \qquad (2^{32})$$
$$a = 3125 \qquad (4 \times 781 + 1)$$
$$c = 907633385$$

Unfortunately, implementing such a congruential algorithm in a high-level language is not as simple as it might appear. The problem lies in the size of the random integers x_i. When a new random number is generated the intermediate result before performing the modulus operation can reach a value of $a(m-1) + c \approx 1 \cdot 3 \times 10^{13}$. Few, if any, implementations of Modula-2 allow integers or cardinals of this size. In fact a cardinal range of 0–65535 is more common. To overcome these difficulties x_i is stored as four cardinals, each covering a range of 0–255. If the four cardinals are represented by s_1, s_2, s_3, and s_4, then

$$x_i = 256^3 s_4 + 256^2 s_3 + 256 s_2 + s_1$$

Operations are performed on $s_1 \ldots s_4$ modulo 256 and overflow during arithmetic operations is therefore avoided. A procedure rnd for generating congruential pseudo-random numbers in the range 0–1 is shown below:

```
VAR
    s1, s2, s3, s4: CARDINAL;

PROCEDURE rnd(): REAL;
CONST
    m = 256;
    a1 = 53; a2 = 12;
    c1 = 233; c2 = 98; c3 = 25; c4 = 54;
VAR
    q: CARDINAL;
    m1: REAL;
```

```
BEGIN
    s1 := a1*s1+c1; q := s1 DIV m; s1 := s1-q*m;
    s2 := a1*s2+a2*s1+c2+q; q := s2 DIV m; s2 := s2-q*m;
    s3 := a1*s3+a2*s2+c3+q; q := s3 DIV m; s3 := s3-q*m;
    s4 := (a1*s4+a2*s3+c4+q) MOD m;
    m1 := FLOAT(m);
    RETURN
        (FLOAT(s4)+(FLOAT(s3)+(FLOAT(s2)+FLOAT(s1)/m1)/m1)/m1)/m1;
END rnd;
```

The random numbers generated by this procedure were tested for uniformity and for correlation using the poker test. A total of one hundred million numbers were generated (on a mainframe computer) and were found not to differ significantly (at the 20% probability level) from true random numbers.

Normally the initial value x_0 will be set to a known value before any calls are made to the procedure *rnd*. For example x_0 could be set to

$$x_0 = 2071690107$$

by the assignments:

```
    s1 := 123; s2 := 123; s3 := 123; s4 := 123;
```

Following such an initialization the same sequence of pseudo-random numbers will always be generated. In many cases this is desirable, for example during the development phase of a Monte-Carlo analysis program. On the other hand, no extra information is gained by running a Monte-Carlo analysis program a second time if the same sequence of random numbers is used.

In some situations, therefore, the initial value x_0 should itself be chosen randomly. One way of doing this relies on there being an accessible real-time clock. The clock is read, and the most rapidly changing element of the time (seconds or fractions of a second) is assigned to the most significant of the four cardinals defining x_0. This is illustrated in the procedure *randomise* given below:

```
PROCEDURE randomise;
CONST
    m = 256;
```

```
VAR
    h: Time;
BEGIN
    GetTime(h);
    s1 := h.minute DIV m;
    s2 := h.minute MOD m;
    s3 := h.millisec DIV m;
    s4 := h.millisec MOD m
END randomise;
```

Any uniform distribution can be derived from the output of the procedure *rnd* by appropriate shifting and scaling. For example, the procedure *uniform* shown below generates a uniform distribution centred on x_0 and of width $\pm dx$.

```
PROCEDURE uniform(x0, dx: REAL): REAL;
BEGIN
    RETURN x0+dx*(2.0*rnd()-1.0)
END uniform;
```

Perhaps the main disadvantage of congruential pseudo-random number generators is that when implemented in a high-level language they are not particularly efficient, involving as they do a number of integer and floating-point operations for each number generated. Normally, however, this disadvantage is outweighed by the fact that such generators are portable and their operation is well understood.

10.7 Non-uniform Distributions

In most cases the distributions of values of electronic components are non-uniform, that is the probability density function $p(x)$ is not a constant. Non-uniform distributions are obtained by performing transformations on the output of a random number generator with a uniform distribution. Two general transformation techniques, which can generate any required distribution, will be described. First, however, the special case of the Gaussian distribution will be considered.

The Gaussian distribution is of unique importance in statistical analysis and special techniques have been developed to generate random numbers with this distribution. When several random numbers are averaged, the result will have an approximately Gaussian distribution. This is a consequence of the Central Limit Theorem of statistics. The larger the number of random values that are averaged, the closer the result is to a Gaussian; the distribution of the original random numbers is not important. If n random numbers are averaged, then the variance of the result is $1/n$ times the variance of the original distribution. This obviously provides a convenient way of generating a Gaussian distribution; a random number y can be obtained from

$$y = \frac{1}{n} \sum_{i=1}^{n} x_i \qquad (10.7.1)$$

where x_i are random numbers with a uniform distribution in the range 0 to 1. The mean of the numbers x_i is $\frac{1}{2}$ and the variance is $\frac{1}{12}$. Consequently the distribution of y will be approximately Gaussian with a mean of $\frac{1}{2}$ and a variance of $1/12n$. What size of n should be used? In practice a value of $n = 12$ gives a good compromise between efficiency and closeness of approximation to a Gaussian. This particular value has the added advantage that a variance of unity can be obtained very simply. The random number y given by the formula

$$y = \sum_{i=1}^{12} x_i \qquad (10.7.2)$$

has a mean of 6 and a variance of 1. Values of y outside the range 0 to 12 will never occur. Although a true Gaussian distribution is not similarly truncated, the probability of a value falling outside the range is extremely small (about 2 in 10^9) and for many purposes this difference is not significant.

A procedure *gaussian* which uses this technique, together with appropriate scaling, to generate random numbers with a mean x_0 and a standard deviation dx, is shown below:

```
PROCEDURE gaussian(x0, dx: REAL): REAL;
VAR
    k: CARDINAL;
    x: REAL;
BEGIN
    x := 0.0;
    FOR k := 1 TO 12 DO x := x+rnd() END;
    RETURN x0+dx*(x-6.0)
END gaussian;
```

An alternative method for generating random numbers with a Gaussian distribution is the Box–Muller method. Unlike the method described above, the Box–Muller transformation generates an exactly Gaussian distribution. If x_1 and x_2 are two random numbers with a uniform distribution in the range 0 to 1, then the formula given below generates a new random value y_1 with a Gaussian distribution

$$y_1 = \sqrt{[-2 \ln(x_1)]} \cos(2\pi x_2) \tag{10.7.3}$$

In fact two random values can be obtained from x_1 and x_2. A second value y_2, which is independent of y_1, is given by the formula:

$$y_2 = \sqrt{[-2 \ln(x_1)]} \sin(2\pi x_2) \tag{10.7.4}$$

The potential improvement in efficiency is, however, offset by an additional complexity in programming. In practice only one of the random values y_1, y_2 is usually generated.

There are certain drawbacks associated with this otherwise very useful transformation. Calculation of logarithm and sine or cosine functions is computationally expensive. More serious is the requirement that the original pairs of random numbers x_1, x_2 must be uncorrelated: any correlation between x_1 and x_2 leads to a skewing of the resultant distribution.

The procedure *gaussian* shown below generates random numbers with a Gaussian distribution of standard deviation *dx* centred on x_0:

```
PROCEDURE gaussian(x0, dx: REAL): REAL;
BEGIN
    RETURN x0+dx*sqrt(-2.0*ln(rnd()))*sin(2.0*pi*rnd())
END gaussian;
```

A more general technique, which can be used to generate many different types of distribution, is based on the cumulative distribution function $q(x)$. This is defined to be the integral of the probability density function $p(x)$:

$$q(x) = \int_{-\infty}^{x} p(z) \, dz \tag{10.7.5}$$

Since $p(x)$ is always positive, $q(x)$ must increase monotonically from 0 to 1 as x varies from $-\infty$ to ∞. A triangular probability density function, together with its corresponding cumulative distribution function, is shown in figure 10.19.

From the definition of the cumulative distribution function (equation 10.7.5) it is clear that

$$\frac{dq}{dx} = p(x)$$

Figure 10.19
A triangular
distribution

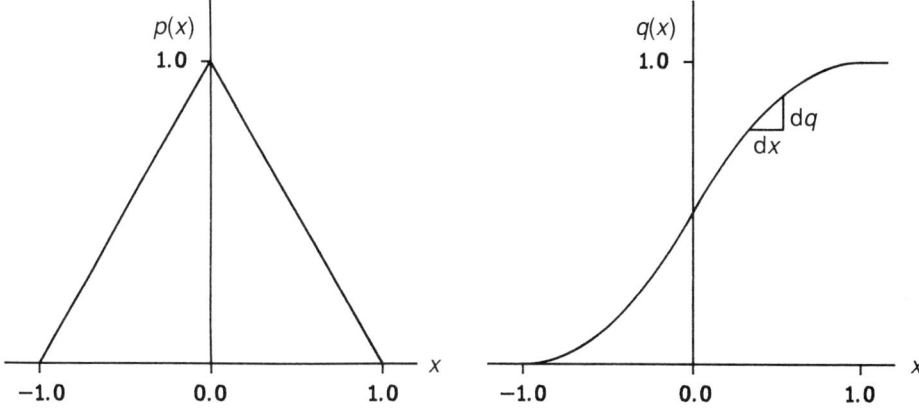

or
$$dq = p(x)\, dx \tag{10.7.6}$$

Suppose that values of q are chosen at random with a uniform distribution in the range 0 to 1. The probability of a number falling within the interval q to $q + dq$ is then simply equal to dq; the corresponding probability of the point falling within the interval dx is $p(x)\, dx$. This is the required distribution. In other words, the appropriate transformation is the inverse of the cumulative distribution function.

Consider the triangular probability density function shown in figure 10.19:

$$
\begin{aligned}
p(x) &= 0 & &\text{for } x < -1 \\
&= 1 + x & &\text{for } -1 \leqslant x < 0 \\
&= 1 - x & &\text{for } 0 \leqslant x < 1 \\
&= 0 & &\text{for } x > 1
\end{aligned}
\tag{10.7.7}
$$

The corresponding cumulative distribution function is

$$
\begin{aligned}
q(x) &= 0 & &\text{for } x < -1 \\
&= 0 \cdot 5 + x + 0 \cdot 5x^2 & &\text{for } -1 \leqslant x < 0 \\
&= 0 \cdot 5 + x - 0 \cdot 5x^2 & &\text{for } 0 \leqslant x < 1 \\
&= 1 & &\text{for } x \geqslant 1
\end{aligned}
\tag{10.7.8}
$$

Finally, the inverse of cumulative distribution function is given by

$$
\begin{aligned}
x &= -1 + \sqrt{(2q)} & &\text{for } q < 0 \cdot 5 \\
&= 1 - \sqrt{(2 - 2q)} & &\text{for } q \geqslant 0 \cdot 5
\end{aligned}
\tag{10.7.9}
$$

The procedure *triangular* shown below uses this method to generate random numbers with the triangular distribution shown in figure 10.19:

```
PROCEDURE triangular(): REAL;
VAR
    q: REAL;
```

```
BEGIN
    q := rnd();
    IF q < 0.5 THEN
        RETURN -1.0+sqrt(2.0*q)
    ELSE
        RETURN 1.0-sqrt(2.0-2.0*q)
    END
END triangular;
```

Two conditions must be met by any probability density function that is to be generated by this method. It must be possible to integrate the function algebraically, and the result of the integration must be capable of being inverted. In practice these are quite serious limitations. For example, a Gaussian cannot be integrated algebraically; the function shown below can be integrated, but not then inverted

$$p(x) = 0 \qquad\qquad \text{for } x < 0$$
$$= \tfrac{1}{2} x^2 e^{-x} \qquad \text{for } 0 \leqslant x < \infty \qquad\qquad (10.7.10)$$

In such cases an alternative technique, the so-called rejection method, can be used.

Suppose that points can be randomly distributed under a probability density function $p(x)$ as shown in figure 10.20. What is the probability that any point has an x coordinate which lies within the interval x to $x + \mathrm{d}x$? A brief consideration of figure 10.20 leads to the conclusion that it is equal to the area of the shaded strip divided by the total area under the curve.

Figure 10.20
Points distributed randomly under a probability density function curve

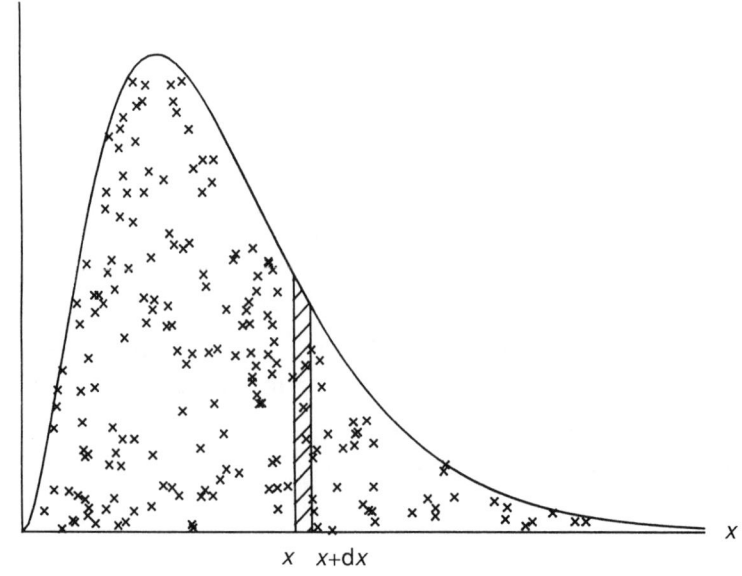

246

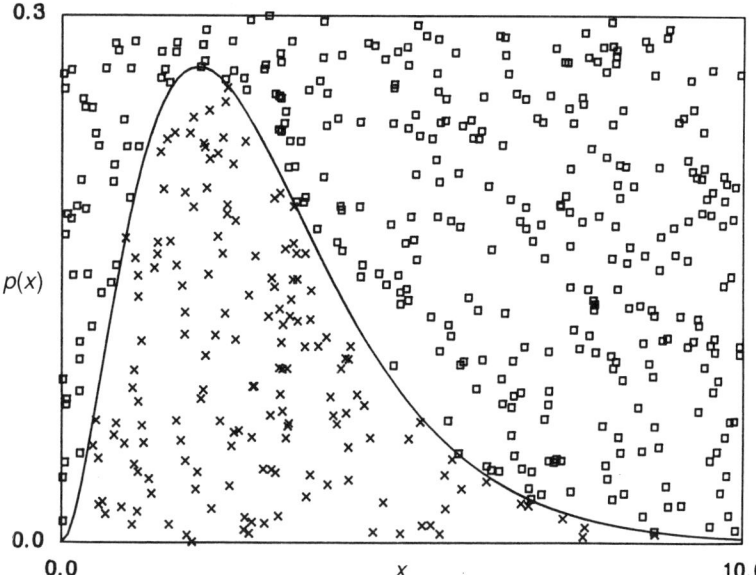

Figure 10.21
Points
distributed
randomly
within a
rectangle

Since the total area is unity, the probability is simply $p(x)\,\mathrm{d}x$. In other words, the x coordinates of the points under the curve have the required probability density function $p(x)$.

There is no straightforward method for directly generating random points under a specified curve. However, it is a simple matter to generate points randomly within a rectangular area containing the curve. A random number generator with a uniform distribution is used to generate pairs of independent x and y coordinates. This is illustrated in figure 10.21. If a point falls above the curve then it is rejected; if it falls below the curve then its x coodinate is returned as the random number with the required distribution.

A procedure *reject* which uses this method to generate random numbers with a probability density function $p(x)$ is shown below. The parameters xm and ym define the size of the rectangle to be used.

```
PROCEDURE p(x: REAL): REAL;
BEGIN
    RETURN 0.5*x*x*exp(-x)
END p;


PROCEDURE reject(p: function; xm, ym: REAL): REAL;
VAR
    x, y, pdf: REAL;
```

```
BEGIN
    REPEAT
        x := xm*rnd( ); y := ym*rnd( );
        pdf := p(x);
        IF (pdf < 0.0) OR (pdf > ym) THEN HALT END;
    UNTIL y<pdf;
    RETURN x
END reject;
```

The efficiency of the rejection method is inversely proportional to the area of the rectangle (since the area under the probability density curve is unity). To achieve the best efficiency, therefore, the smallest rectangle that contains the probability density function should be used. A test is included in the procedure *reject* to halt the program if the rectangle is found to be too small.

One obvious limitation of the rejection method is that it cannot deal with probability density functions which extend over an infinite range. In many cases, however, it is possible to truncate such functions without serious effect. The function given in equation (10.7.10) is such a case. As a consequence of its exponential approach to zero, only $0 \cdot 5\%$ of the area of its function lies above $x = 10$, and for most purposes it can be truncated at this point.

10.8 Summary

Electronic components have values which lie in a tolerance band around their nominal values and when an electronic system is manufactured from such components it may fail to meet its specification. Tolerance analysis is used to assess the effects of component variations on the performance of electronic systems. There are two types of tolerance analysis: worst-case and statistical analysis.

Worst-case analysis seeks to determine whether any combination of component values within their tolerance ranges leads to a performance that violates the specification. The simplest type of worst-case analysis assumes that the performance functions making up the specification vary linearly with the n component values. Only $n + 1$ evaluations of the performance functions are then required to determine the worst case. If, on the other hand, linearity cannot be assumed then all combinations of component value extremes must be investigated. This is known as vertex analysis and involves 2^n evaluations of the performance functions. Vertex analysis is therefore impracticable for systems containing more than about 16 components.

Worst-case tolerance analysis does not attempt to determine manufacturing yield if it is less than 100%. It tends to be pessimistic, predicting a failure to meet the specification even if the failure probability is very small.

Statistical tolerance analysis methods make use of the component value distributions to calculate manufacturing yields. Components which have not been subject to selection by the manufacturer will usually have an approximately Gaussian distribution of values. Selection will lead to truncation of the Gaussian distribution, and if a narrow band of values is selected then its distribution will be nearly uniform.

The method of moments is a statistical analysis technique in which the mean and variance of a performance function are determined from the means and variances of the component values. Since the distribution of a performance function is expected to be approximately Gaussian it is completely determined by its mean and variance. The manufacturing yield can be estimated from the performance function distribution and the specification limits. Although the method of moments is efficient and straightforward, the approximations used to derive the mean and variance of the performance function are only valid for small component tolerances.

Monte-Carlo analysis attempts to simulate the manufacturing process on a computer by selecting component values randomly according to their known distributions. For each set of component values the performance functions are evaluated and compared with their specification limits. The process is repeated a large number of times and the proportion of failures is used to estimate the true failure rate. Monte-Carlo tolerance analysis is accurate and reliable, but tends to involve a substantial amount of computation.

The heart of Monte-Carlo analysis is the random number generator. True random numbers cannot be generated by a deterministic machine and in practice pseudo-random numbers are used instead. Pseudo-random numbers are normally generated initially with a uniform distribution in the interval 0 to 1. The mixed congruential algorithm is suitable for implementation in a high-level language and produces pseudo-random numbers that are for most practical purposes indistinguishable from true random numbers.

Random numbers with a uniform distribution can be converted to a Gaussian distribution in two ways. The sum of several uniform random numbers will have an approximately Gaussian distribution. Alternatively the Box–Muller transformation generates numbers with an exactly Gaussian distribution from pairs of uniform random numbers. Other types of non-uniform distribution can be produced either by inverting the cumulative distribution function or, if this is not possible, by the rejection method.

10.9 Problems

1 Figure 10.22 shows the network diagram of a twin-T filter. The component values have been chosen to give a frequency response notch at 50 Hz. Resistors with a uniform distribution over a tolerance range of $\pm 2\%$ and capacitors with a uniform distribution over a tolerance range of $\pm 1\%$ are used to construct the filter. Use vertex analysis to determine the worst-case (that is the highest) value of the gain at 50 Hz.

The specification calls for the gain at 50 Hz to be less than -40 dB. Perform a 10000-trial Monte-Carlo analysis to determine the maximum value of the true failure rate at the 97% confidence level.

Figure 10.22
A twin-T filter

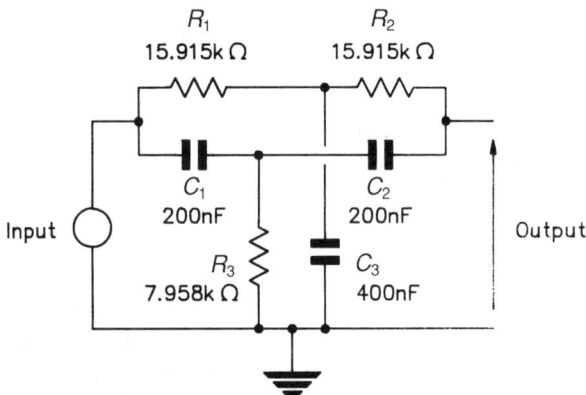

R_1 15.915k Ω
R_2 15.915k Ω
C_1 200nF
C_2 200nF
Input
R_3 7.958k Ω
C_3 400nF
Output

2 The third-order low-pass equally-terminated ladder filter shown in figure 10.23 has a Chebychev response with 3 dB passband ripple and a cut-off frequency of 1 kHz. Components with Gaussian distributions of values are used to construct this filter. The standard deviations for the resistors are equal to 5% of their nominal values and the standard deviations for the reactive components are equal to 2% of their nominal values.

At low frequencies the gain g_0 of the filter is $-6 \cdot 02$ dB. The specification requires that the gain at 800 Hz must lie between $g_0 - 1 \cdot 5$ dB and

Figure 10.23
A passive third-order low-pass filter

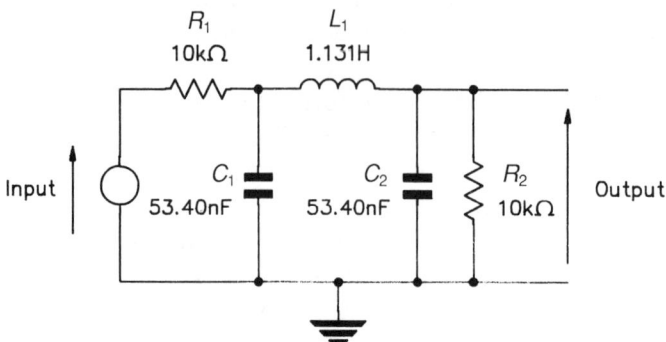

R_1 10kΩ
L_1 1.131H
C_1 53.40nF
C_2 53.40nF
R_2 10kΩ
Input
Output

250

$g_0 + 0 \cdot 5$ dB. Use the method of moments to estimate the failure rate. Perform a 10000-trial Monte-Carlo analysis to determine the maximum value of the true failure rate at the 97% confidence level.

3   The active filter shown in figure 10.24 is a cascade implementation of a Chebychev response with 3 dB passband ripple and a cut-off frequency of 1 kHz. Components with Gaussian distributions of values are used to construct this filter. The standard deviations for the resistors are equal to 5% of their nominal values and the standard deviations for the capacitors are equal to 2% of their nominal values.

At low frequencies the gain g_0 of the filter is 0 dB. The specification requires that the gain must lie between $g_0 - 1 \cdot 5$ dB and $g_0 + 0 \cdot 5$ dB. Use the method of moments to estimate the failure rate. Perform a 10000-trial Monte-Carlo analysis to determine the maximum value of the true failure rate at the 97% confidence level. Compare these results with those of the passive filter in the previous problem.

Figure 10.24
A third-order low-pass active filter

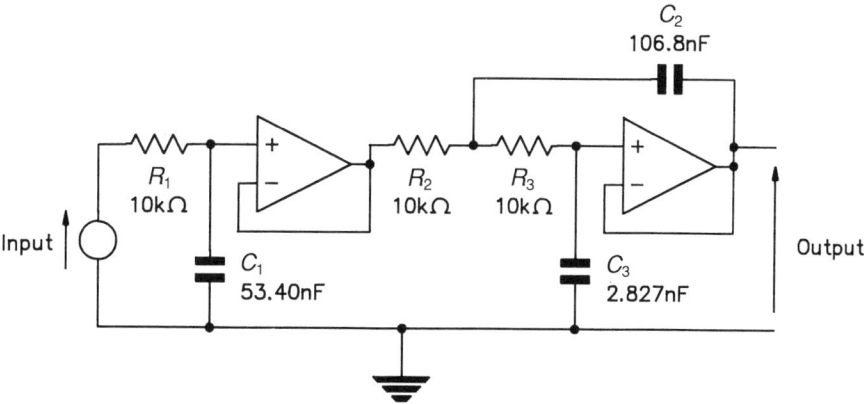

4   A useful empirical method for evaluating the output of a pseudo-random number generator is the poker test. Groups of five random numbers in the range 0 to 1 are generated, and each number is placed in one of the ten subranges 0 to $0 \cdot 1$, $0 \cdot 1$ to $0 \cdot 2$, ... $0 \cdot 9$ to $1 \cdot 0$. A group is categorized according to the largest content k of any subrange; k can have values, 1, 2, 3, 4 or 5.

Perform a poker test on 100000 pseudo-random numbers generated by the mixed congruential algorithm, and determine the frequencies of occurrence p_k of the five possible values of k. Compare these with the ideal frequencies q_k:

$q_1 = 0 \cdot 3024$
$q_2 = 0 \cdot 6120$
$q_3 = 0 \cdot 0810$
$q_4 = 0 \cdot 0045$
$q_5 = 0 \cdot 0001$

251

Obtain the value of χ^2 from

$$\chi^2 = \sum_{i=1}^{n} \frac{(p_i - q_i)^2}{q_i}$$

If χ^2 is less than $9 \cdot 5$ then the pseudo-random numbers can be regarded as indistinguishable from true random numbers according to the poker test. There is only a 5% probability that χ^2 will exceed $9 \cdot 5$, and a 1% probability that χ^2 will exceed $11 \cdot 1$ by pure chance.

Investigate the effects of changing the multiplier a and the increment c in the mixed congruential algorithm.

5 Random numbers are required with a Lorentzian probability density function $p(x)$:

$$p(x) = \frac{1}{\pi(1 + x^2)}$$

Obtain the cumulative probability function $q(x)$ that corresponds to $p(x)$. Hence derive a transformation that will generate random numbers with a Lorentzian probability density function from random numbers with a uniform distribution.

Using this transformation generate 1000 random numbers with the Lorentzian distribution. Plot a histogram showing the occurrence of these random numbers over the range -10 to 10.

Repeat this exercise using the rejection method. Compare the efficiencies of the two methods for generating Lorentzian random numbers.

Appendix A: SPICE Network Simulator

A.1 Introduction

The most widely available computer program for the analysis of electronic networks must surely be SPICE. This program is essentially a network simulator which, given the description of a network, predicts its behaviour under various conditions. SPICE was originally developed at the Electronic Research Laboratory of the University of California, and was released in 1975 as a Fortran program for running in batch mode on mainframe computers. Recent implementations of SPICE on personal computers tend to be more interactive than the original program; in particular the output data is often presented in high-resolution graphical form. The mainframe origins of SPICE are, however, still apparent from the text format of the input data.

In common with other computer-aided network analysis techniques, SPICE is most useful in the later stages of the design process. It can be used to confirm the correct operation of a network, or to determine the effects of varying component values, more quickly and cheaply than by actually constructing the network. In the early stages of a design, when decisions concerning alternative network topologies are being made, it is of limited value.

Networks which are to be analyzed by SPICE may contain the following elements:

Resistors, capacitors and inductors
Mutual inductors
Independent voltage and current sources
Controlled sources
Transmission lines
Semiconductor diodes
Bipolar junction transistors
Junction field-effect transistors
MOS field-effect transistors

SPICE provides for three different types of analysis: non-linear d.c. analysis, non-linear transient analysis and small-signal a.c. analysis. In d.c. analysis the operating point of the network is determined by solving the non-linear equations governing the behaviour of the network with the capacitors open-circuited and the inductors short-circuited. By performing d.c. analysis for a sequence of voltages or currents at an independent source, the d.c. transfer characteristic of the network can be obtained. A d.c. analysis is automatically performed prior to transient analysis in order to establish the initial conditions. It is also performed prior to small-signal a.c. analysis in order to determine the operating points, and therefore the small-signal models, of any non-linear elements.

SPICE performs transient analysis with a variety of input waveforms over a time interval specified by the user. Large-signal models are used to represent non-linear devices such as diodes and transistors during transient analysis. Input waveforms recognized by SPICE include pulses of specified width, height and risetime, and sinusoids of specified amplitude and frequency. An option is available for Fourier analysis of the output waveforms.

Small-signal a.c. analysis is performed by SPICE using linearised small-signal models of the non-linear devices. The outputs resulting from one or more a.c. inputs are calculated over a frequency range specified by the user. SPICE will also calculate the voltage or current at any point in the network that results from noise generated in the resistors and semiconductor devices.

The semiconductor models used by SPICE have, of course, a strong temperature dependence. Other elements such as resistors are normally assumed to be temperature-independent, but if required can be given a suitable temperature coefficient. SPICE allows the operating temperature to be set, but if no temperature is specified then a default value of $27^{\circ}C$ is used.

An iterative process is employed in both d.c. and transient analysis to obtain solutions to the non-linear equations. Iteration is normally stopped when the node voltages have converged to within $0 \cdot 1\%$ or $1\ \mu V$, whichever is the larger, and the branch currents have converged to within $0 \cdot 1\%$ or 1 pA, whichever is the larger. Although the algorithm used by SPICE is generally very reliable, occasionally it may fail to converge. When this happens the program terminates after printing out the node voltages at the last iteration. Special precautions must be taken to ensure convergence during d.c. analysis of networks which have positive feedback.

A.2 Network Description

SPICE uses a free format for the input data with each line normally containing a separate item. Fields within a line are separated by delimiters which can be spaces, commas, equal signs or parentheses; extra spaces following a delimiter are ignored. When it is necessary to continue an item onto a further line, a continuation sign, consisting of a plus sign in the first column of the new line, can be used. Any line starting with an asterisk is treated as a comment and is ignored by SPICE. However, comments are very important and should be used liberally to aid future understanding.

The first line of input data must be a network title, and the last line must be an end statement (.END). In between the order is arbitrary, except that continuation lines must immediately follow the lines being continued. Each network element is specified by an element line which consists of the element name, followed by the nodes to which the element is connected and the values of any parameters required to define the electrical characteristics of the element. For example the line shown below might be used to specify a 120 Ω resistor connected between node 4 and ground:

```
RLOAD   4   0   120OHM
```

Element names must start with a letter, which may be followed by any combination of letters and digits. Only the first eight characters in a name are significant, however. Thus R42, ZOUT, UA741, BC184K, FEEDBACKRESISTOR and FEEDBACKCAPACITOR are all valid names, but the last two are not distinguished by SPICE. The first letter of an element name designates the element type so that, for example, names starting with the letter R represent resistors. In element descriptions the notation Z******* will be used for a valid name, of any length, which starts with the letter Z.

Nodes are specified by integer numbers. Each node of the network must be assigned a number, with the number 0 being reserved for the ground or reference node. Node numbers need not be consecutive.

Parameters, which define the electrical characteristics of the elements, are specified by floating-point numbers. Exponential notation may be used and the following are all correctly-formed floating-point numbers: 99, -3, $2 \cdot 7182818$, $1 \cdot 0E-9$ and $-1 \cdot 414E3$. Numbers may be followed by symbols representing scaling factors. SPICE recognizes the following scaling factors:

Symbol	Factor
F	10^{-15}
P	10^{-12}
N	10^{-9}
U	10^{-6}
M	10^{-3}
K	10^{3}
MEG	10^{6}
G	10^{9}
T	10^{12}

With the exception of these symbols, any letter following a number or scaling factor is ignored. Hence 5, 5A, 5AMP, 5HZ and $0 \cdot 005$KHZ all represent the same number. Some parameters are optional; in the element descriptions which follow optional parameters will be enclosed in triangular brackets ⟨ ⟩.

A.3 Resistors, Capacitors and Inductors

The general forms of element descriptions for resistors, capacitors and inductors are given below:

```
R*******   n1   n2   value   <TC=tc1<,tc2>>
C*******   n1   n2   value   <IC=vinit>
L*******   n1   n2   value   <IC=cinit>
```

The integers $n1$ and $n2$ specify the two nodes to which the element is connected, and *value* is a floating-point number giving the resistance in ohms, capacitance in farads or the inductance in henrys.

Resistors can have optional temperature coefficient parameters. The resistance r is assumed to vary according to:

$$r(\theta_1) = r(\theta_0)\{1 + tc1(\theta_1 - \theta_0) + tc2(\theta_1 - \theta_0)^2\}$$

where θ_0 is the nominal temperature (27°C) and θ_1 is the network temperature. If not otherwise specified, the temperature coefficients $tc1$ and $tc2$ are taken to be zero. A 10 kΩ resistor with a temperature coefficient of $-0 \cdot 1\%$ connected bctween nodes 4 and 5 could be described by

```
RIN   4   5   10KOHM   TC=-0.001
```

Inductors and capacitors can have an optional parameter which defines the initial conditions; *vinit* specifies the initial voltage across a capacitor and *cinit* specifies the initial current through an inductor. These initial conditions are only used if the *UIC* option is active, otherwise the initial conditions are established by d.c. analysis of the network.

Consider the passive ladder filter shown in figure A.1. The passive

Figure A.1
A third-order
low-pass
elliptic filter

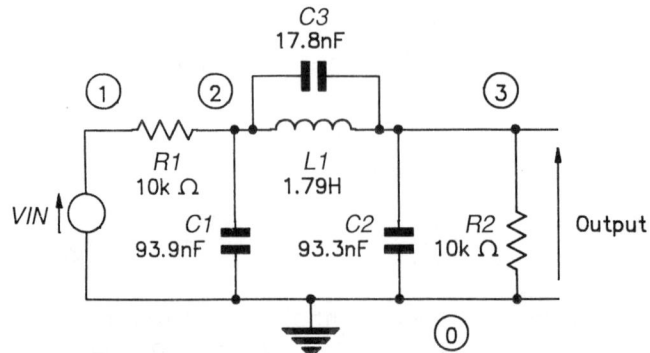

256

elements of this network could be represented by

```
R1   1   2   10KOHM
R2   3   0   10KOHM
C1   2   0   93.31NF
C2   3   0   93.31NF
C3   2   3   17.78NF
L1   2   3   1.789H
```

This does not, however, constitute a complete SPICE network description. In particular, no input generator has been defined, nor has the type of analysis been specified.

A.4 Independent Sources

Independent sources are used primarily for signal inputs, for power supplies and as dummy voltage sources for measuring branch currents. Both voltage and current independent sources are recognized by SPICE and the general form of their element descriptions are given below:

```
V*******   n1   n2   <<DC>dcval>   <AC<acmag<acphase>>>   <trfunc>
I*******   n1   n2   <<DC>dcval>   <AC<acmag<acphase>>>   <trfunc>
```

The integers *n1* and *n2* specify the nodes to which the source is connected. The sign convention is that for a voltage source *n1* is positive with respect to *n2*, and for a current source the current flows from *n1* through the source to *n2*.

If the optional parameters are omitted then the value of the source is assumed to be zero. Voltage sources of zero value are, of course, equivalent to short circuits, and can therefore be inserted into a network without affecting its performance. This might at first appear to be a pointless exercise. However, SPICE determines the current that flows in all of the independent voltage sources, and dummy sources can therefore be used to probe the currents in any branch of the network.

Time-invariant source values (representing, for example, d.c. power supplies) that are to be used during d.c. analysis are specified by the *dcval* parameter, optionally preceded by the keyword *DC*. For example, the two independent source definitions given below are alternative ways of representing a 5 V d.c. power supply connected between ground and node 2:

```
VS   2   0   5V
VS   2   0   DC 5V
```

Time-dependent source values to be used during transient analysis can be specified by the use of the *trfunc* parameter. If a source is assigned a time-dependent value, then the zero-time value of the source function is used for d.c. analysis. SPICE recognizes five different source functions: pulse, sinusoid, exponential, piecewise linear and single-frequency FM. The use of time-dependent source values is described in detail in section A.9 which deals with transient analysis.

Independent source values which are to be used during a.c. analysis must include the keyword *AC* in their specification. The magnitude of the a.c. signal can be specified by the optional *acmag* parameter; if it is not specified then a magnitude of unity is assumed by default. The phase can also be specified by the optional *acphase* parameter; the default value for this parameter is zero.

A suitable independent source definition for the input generator in figure A.1 would be

```
VIN  1  0  AC  1V
```

Since there is only one input to the network it is unnecessary to specify the phase.

A.5 A.C. Analysis

Before a.c. analysis is performed the d.c. operating point is automatically determined. This allows the linearized small-signal models for any non-linear devices to be established. Then the frequency is stepped through a range of values, and at each frequency the responses of the network nodes to the a.c. independent sources are calculated. Note that the result of a.c. analysis will be meaningless unless at least one independent source definition includes the keyword *AC*. A frequency sweep statement is used to specify the analysis frequencies:

```
.AC  type  np  fstart  fstop
```

The initial frequency is *fstart* and the final frequency is *fstop*. A linear frequency sweep is obtained by setting *type* to *LIN*; *np* is then the total number of analysis frequencies. If a logarithmic sweep is required then *type* should be set to *OCT* or *DEC*; in this case, *np* is respectively the number of analysis frequencies per octave or the number of analysis frequencies per decade. For example, the statement below calls for a logarithmic sweep from 30 Hz to 3 kHz with a total of 16 analysis frequencies:

```
.AC  DEC  8  30HZ  3KHZ
```

SPICE allows the results of an analysis to be presented either in tabular form, or as a crude graph made up of printable ASCII characters. A print statement is used to request tabular output and has the general form:

```
.PRINT  prtype  ov1  <ov2  ..  <ov8>>
```

The parameter *prtype* specifies the type of analysis to be displayed and can be set to *AC*, *DC*, or *TRAN*, representing a.c., d.c. and transient analysis respectively.

Up to 8 output variables, represented by *ov1 ... ov8* in the print statement, can be printed at each point in the analysis. Voltage output specifications have the general form:

```
V(n1<,n2>)
```

The voltage difference between nodes *n1* and *n2* is printed. If the second node *n2* is omitted then the reference node 0 is used by default. Current output specifications have the general form:

```
I(V*******)
```

The current flowing in the independent voltage source *V******* is printed, using the convention that a positive current flows into the positive node, and out of the negative node of the source.

In a.c. analysis these output specifications cause the magnitude of the a.c. voltages or currents to be printed. Other properties of the outputs, such as the real part or the phase, can be obtained by the use of the following output specifications (which may only be used during a.c. analysis):

Property	Voltage	Current
Real part	VR(n1<,n2>)	IR(V*******)
Imaginary part	VI(n1<,n2>)	II(V*******)
Magnitude	VM(n1<,n2>)	IM(V*******)
Phase	VP(n1<,n2>)	IP(V*******)
Decibels	VDB(n1<,n2>)	IDB(V*******)

For example, the statement given below prints the magnitude in dB and the phase of the a.c. voltage at node 3:

```
.PRINT  AC  VDB(3)  VP(3)
```

A complete SPICE description of the network shown in figure A.1, with network title, network elements, analysis statement, print statement and end statement, would be

```
Elliptic Filter
* Third-order Elliptic filter with 3db passband ripple
* Implemented as an equally-terminated ladder filter
R1  1  2  10KOHM
R2  3  0  10KOHM
C1  2  0  93.31NF
C2  3  0  93.31NF
C3  2  3  17.78NF
```

```
L1   2   3   1.789H
VIN   1   0   AC   1V
.AC   DEC   8   30HZ   3KHZ
.PRINT   AC   VDB(3)   VP(3)
.END
```

SPICE will process this data and generate output in three sections. The first section is simply a copy of the network description. The second section, which is shown below, consists of a d.c. analysis to establish the operating point of any non-linear devices. In the case of this passive filter which has no external power sources the d.c. currents and voltages are all zero.

```
******** 01/16/90 *********** Spice   *********** 20:23:09 ********
Elliptic Filter
****   SMALL SIGNAL BIAS SOLUTION      TEMPERATURE =   27.000 DEG C
*****************************************************************


NODE VOLTAGE      NODE VOLTAGE      NODE VOLTAGE      NODE VOLTAGE
(  1)  0.0000      (  2)  0.0000      (  3)  0.0000


VOLTAGE SOURCE CURRENTS
    NAME           CURRENT
    VIN            0.000E+00


    TOTAL POWER DISSIPATION   0.00E+00   WATTS
```

Finally SPICE prints the output data in tabular form:

```
******** 01/16/90 *********** Spice   *********** 20:23:09 ********
Elliptic Filter
****   AC ANALYSIS                     TEMPERATURE =   27.000 DEG C
*****************************************************************
    FREQ        VDB(3)        VP(3)
    3.000E+01   -6.128E+00   -1.095E+01
    4.001E+01   -6.209E+00   -1.450E+01
    5.335E+01   -6.348E+00   -1.911E+01
    7.114E+01   -6.577E+00   -2.498E+01
    9.487E+01   -6.940E+00   -3.226E+01
    1.265E+02   -7.469E+00   -4.095E+01
    1.687E+02   -8.146E+00   -5.090E+01
```

```
2.250E+02    -8.810E+00    -6.216E+01
3.000E+02    -8.947E+00    -7.632E+01
4.001E+02    -7.046E+00    -1.065E+02
5.335E+02    -1.267E+01     1.532E+02
7.114E+02    -3.013E+01     1.205E+02
9.487E+02    -4.872E+01    -7.004E+01
1.265E+03    -3.944E+01    -7.587E+01
1.687E+03    -3.933E+01    -7.969E+01
2.250E+03    -4.075E+01    -8.238E+01
3.000E+03    -4.272E+01    -8.433E+01
```

An alternative form of output can be obtained by the use of a plot statement. This generates a crude graphical display using printable ASCII characters and has the general form:

```
.PLOT  pltype  ov1<(lo1,hi1)>  <ov2<(lo2,hi2)> .. <ov8>>
```

The parameter *pltype* specifies the type of analysis to be displayed, and can be set to *DC*, *AC* or *TRAN*. Up to 8 output variables, represented by *ov1 ... ov8* in the plot statement, can be plotted at each point in the analysis. The format of the output variables is the same as for print statements, except that optional upper and lower plotting limits (*lo, hi*) may be specified; if these are omitted then SPICE automatically determines the maximum and minimum values of all the output variables, and uses these as the plotting limits. The plot statement shown below will display the magnitude in dB and the phase of the voltage at node 3; the limits on the magnitude are set to -60 dB and 0 dB, and the limits on the phase are set to $-180°$ and $+180°$.

```
.PLOT  AC  VDB(3)(-60,0)  VP(3)(-180,180)
```

The frequency response of the elliptic filter obtained using this plot statement is shown on page 262.

Although this graphical output may be preferable to tabular output for some purposes, its resolution is very poor, and if more than one variable is displayed the result may be confusing. This is the case in the example shown on page 262, where the gain and phase plots overlap at frequencies above 500 Hz. Fortunately most modern SPICE implementations for personal computers or workstations now include a graphics post-processor. When such SPICE programs are run they generate, in addition to the standard output, a graphics datafile. This file can be interpreted by the post-processor to produce a high-resolution graphical display and also, if required, hard copy on a printer or plotter.

Figure A.2 shows a display of the gain and phase of the elliptic filter as generated by a typical graphics post-processor. The advantages of

```
******** 01/16/90 *********** Spice    *********** 20:23:09 ********
Elliptic Filter
****  AC ANALYSIS                       TEMPERATURE =   27.000 DEG C
********************************************************************
LEGEND:
*: VDB(3)
+: VP(3)
(*) -6.0E+01        -4.5E+01         -3.0E+01         -1.5E+01        0.0E+00
(+) -1.8E+02        -9.0E+01          0.0E+00          9.0E+01        1.8E+02

          - - - - - - - - - - - - - - - - - - - - - - - - - -
3.00E+01 .              .              + .              .         *         .
3.63E+01 .              .              + .              .         *         .
4.40E+01 .              .              + .              .         *         .
5.33E+01 .              .             +  .              .         *         .
6.46E+01 .              .            +   .              .         *         .
7.83E+01 .              .            +   .              .         *         .
9.48E+01 .              .           +    .              .         *         .
1.14E+02 .              .         +      .              .        *          .
1.39E+02 .              .        +       .              .        *          .
1.68E+02 .              .      +         .              .       *           .
2.04E+02 .              .     +          .              .       *           .
2.47E+02 .              .    +           .              .       *           .
3.00E+02 .              . +              .              .       *           .
3.63E+02 .              +                .              .       *           .
4.40E+02 .         +    .                .              .        *          .
5.33E+02 .              .                .              . *           +     .
6.46E+02 .              .                .         *    .        +          .
7.83E+02 .              .      *         .              .    +              .
9.48E+02 .         *    .  +             .              .                   .
1.14E+03 .              . + *            .              .                   .
1.39E+03 .              . +   *          .              .                   .
1.68E+03 .              . +  *           .              .                   .
2.04E+03 .              .+  *            .              .                   .
2.47E+03 .              .+ *             .              .                   .
3.00E+03 .              .+*              .              .                   .
          - - - - - - - - - - - - - - - - - - - - - - - - - -
```

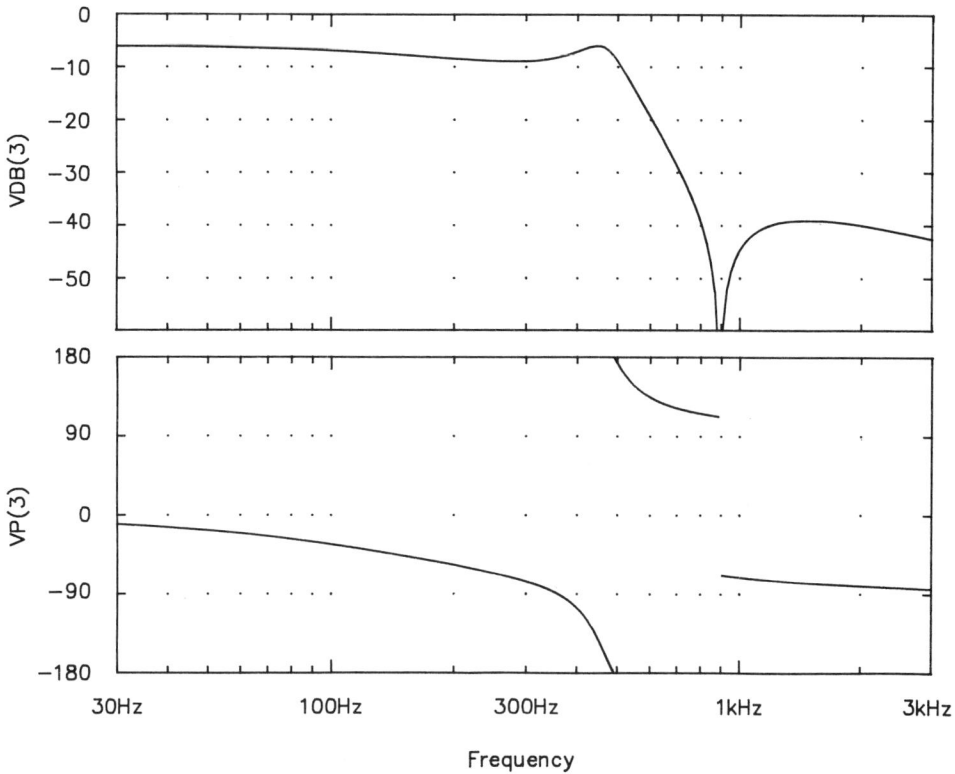

Figure A.2 Gain and phase of the elliptic filter

presenting SPICE output in this way are obvious, and high-resolution graphical displays will be used wherever possible in the examples which follow.

A.6 Semiconductor Devices

SPICE recognizes four types of semiconductor device, namely junction diodes, bipolar junction transistors (BJTs), junction field-effect transistors (JFETs) and MOSFETs. Linear passive devices such as resistors and capacitors are completely specified by a few parameters. Semiconductor devices, by contrast, have complex characteristics which involve many parameters. Moreover a network will often contain several semiconductor devices with similar characteristics. In order to avoid having to specify the parameters for each device, semiconductor element descriptions include a reference to a previously defined device model.

Each semiconductor element description contains the device name, the nodes to which it is connected and the model name. Additional optional parameters may also be included. The general forms of element descriptions for junction diodes, BJTs, JFETs and MOSFETS are respectively:

```
D******* na  nc  mname  <AREA=aval>  <OFF>  <IC=vd>

Q******* nc  nb  ne  mname  <AREA=aval>  <OFF>  <IC=vbc,vce>

J******* nd  ng  ns  mname  <AREA=aval>  <OFF>  <IC=vds,vgs>

M******* nd  ng  ns  nb  mname  <OFF>  <IC=vds,vgs,vbs>
```

Junction diodes have two connection nodes: *na* is the anode and *nc* is the cathode. The diode model to be used is specified by the model name *mname*. *AREA* is an optional parameter which determines the equivalent number of parallel devices; if it is omitted then a value of $1 \cdot 0$ is assumed. *OFF* sets the starting conditions to off for this element in d.c. analysis, and *IC*, which is only active if the *UIC* option is selected, sets the initial conditions to be used in transient anlaysis. An example of an element description for a diode is

```
DBRIDGE  2  4  D1N4148
```

Bipolar junction transistors have three connection nodes: *nc* is the collector, *nb* the base and *ne* the emitter. The effect of the optional parameters is the same as for junction diodes. Junction field-effect transistors and MOSFETS have three main connection nodes: *nd* is the drain, *ng* is the gate and *ns* is the source. In the case of MOSFETS there is an additional node *nb* which is the substrate connection.

A.7 Semiconductor Models

Semiconductor device models must be defined either within the network description or in a separate model library. In the latter case a request must be made to include the appropriate library file:

```
.LIB  mylib.cir
```

Device models are defined by model statements which have the general form:

```
.MODEL  mname  type  (<P1=val1>  <P2=val2> .. )
```

The model name is *mname*, and *type* specifies which of the possible semiconductor devices are being modelled:

Type	Description
D	diode model
NPN	NPN bipolar transistor model
PNP	PNP bipolar transistor model
NJF	N-channel JFET model
PJF	P-channel JFET model
NMOS	N-channel MOSFET model
PMOS	P-channel MOSFET model

Figure A.3
Diode
equivalent
circuit for
transient
analysis

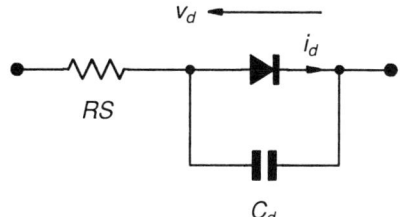

Parameter values $P1, P2, \ldots$ are optional; if they are omitted then default values are used. Each type of semiconductor device has its own set of parameters. For example, SPICE uses the diode equivalent circuit shown in figure A.3 for the purpose of d.c. and transient analysis.

An ohmic resistance RS is in series with an ideal diode whose characteristic in the forward region is defined by the saturation current IS and the emission coefficent N according to

$$i_d = IS\{\exp(v_d/Nv_t) - 1\} \tag{A.7.1}$$

The thermal voltage $v_t = kT/q$ is equal to approximately 26 mV at room temperature. Both v_t and IS are temperature-dependent.

In the reverse bias region, the diode is characterized by its reverse breakdown voltage BV and the current IBV at the breakdown voltage:

$$i_d = -IBV \exp\{-(v_d + BV)/(Nv_t)\} \tag{A.7.2}$$

The capacitance C_d represents charge storage effects in the diode and has a complex dependency on the applied voltage v_d.

A suitable model for a general-purpose fast diode, such as the IN4148, would be

```
**   1N4148 General purpose fast diode
.MODEL D1N4148  D  (IS=2.0E-9 RS=0.8 N=1.8 TT=4.0E-9
+                   CJO=1.4E-12 BV=100)
```

The parameters TT and CJO are associated with the diode capacitance C_d. No value for the reverse breakdown current IBV is given and the default value of 1 mA will therefore be used.

Similar, although much more complex, models are used for BJTs, JFETs and MOSFETS.

A.8 D.C. Analysis

D.C. analysis determines the operating point of a network with all inductors short-circuited and all capacitors open-circuited. It is automatically performed prior to a.c. and transient analysis, or if no other form of analysis is requested. Consider the non-linear network shown in figure A.4. Networks of this type are used for piecewise-linear shaping and node 7 is

Figure A.4
A non-linear
network

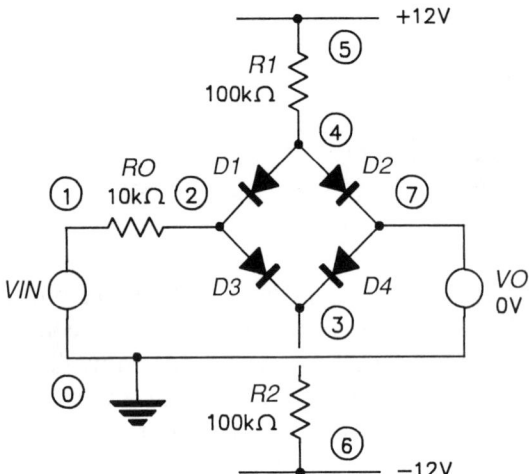

normally connected to the virtual ground of an operational amplifier. The virtual ground has been replaced by a zero-voltage source *VO* which acts as a current probe. A SPICE description of this network is given below:

```
Diode Non-Linear Network
* One element of piecewise-linear network
* Node 7 is a virtual ground
.LIB mylib.cir
RO    1   2   10.0KOHM
D1    4   2   D1N4148
D2    4   7   D1N4148
D3    2   3   D1N4148
D4    7   3   D1N4148
R1    5   4   100.0KOHM
R2    3   6   100.0KOHM
VS1   5   0   DC  12.0V
VS2   0   6   DC  12.0V
VIN   1   0   DC   1.0V
VO    7   0   DC   0.0V
.END
```

The input source *VIN* has been set to 1 V. A SPICE analysis using this network description gives the following results:

```
******* 01/16/90 ********** Spice   ********** 20:29:09 *******
Diode Non-Linear Network
****   SMALL SIGNAL BIAS SOLUTION      TEMPERATURE =   27.000 DEG C
*****************************************************************
```

NODE VOLTAGE	NODE VOLTAGE	NODE VOLTAGE	NODE VOLTAGE
(1) 1.0000	(2) .0970	(3) −.4083	(4) .5048
(5) 12.0000	(6)−12.0000	(7) 0.0000	

VOLTAGE SOURCE CURRENTS

NAME	CURRENT
VS1	−1.150E−04
VS2	−1.159E−04
VIN	−9.030E−05
VO	8.934E−05

TOTAL POWER DISSIPATION   2.86E−03   WATTS

These results indicate that with an input voltage of 1 V the output current is 89·34 μA. To obtain the complete input/output characteristic however, the d.c. analysis must be performed for a sequence of voltages between $-2·5$ V and 2·5 V. It would obviously be very inconvenient to edit the network description and perform a SPICE analysis for each input voltage.

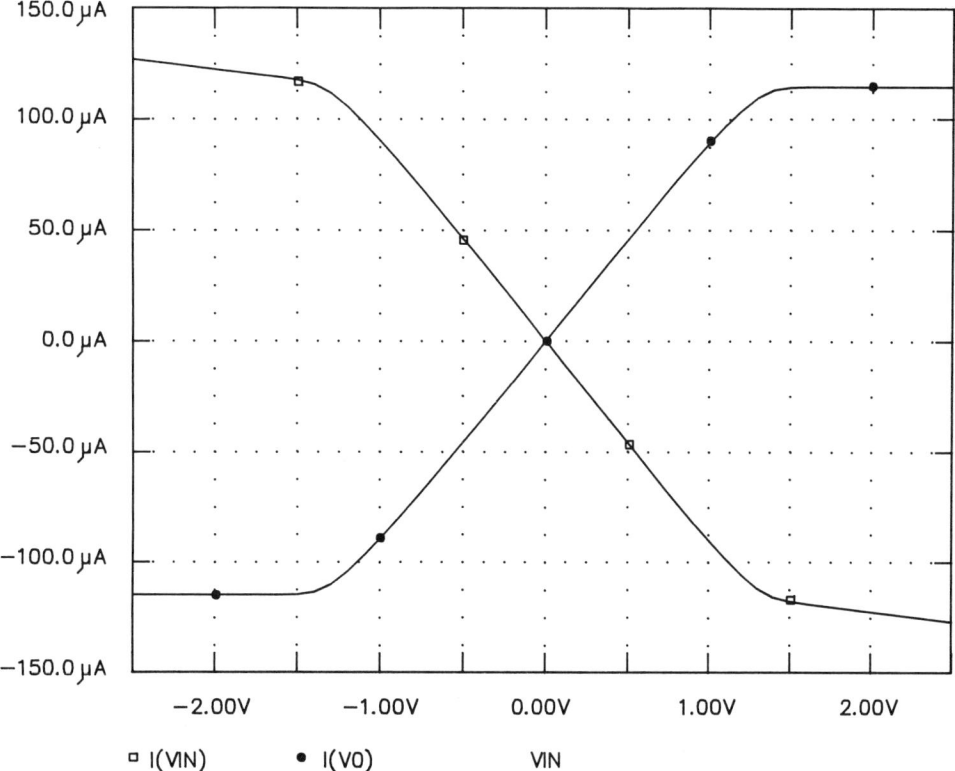

Figure A.5 Input/output characteristic of the non-linear network

Fortunately SPICE allows the voltage or current at an independent source to be swept through a range of values. A d.c. sweep request has the following form:

```
.DC  srcname  start  stop  inc
```

The value of the independent source *srcname* is swept from *start* to *stop* in increments of *inc*. A print or plot statement specifying the output variables must follow the d.c. sweep request.

The extra statements necessary to sweep the input voltage *VIN* of the non-linear network from $-2 \cdot 5$ V to $2 \cdot 5$ V and to plot the input current *I(VIN)* and the output current *I(VO)* are

```
.DC  VIN  -2.5V  2.5V  0.1V
.PRINT  DC  I(VIN)  I(VO)
```

Figure A.5 shows the results of this swept d.c. analysis.

A.9 Transient Analysis

SPICE can be used to compute the values of the output variables as a function of time over a specified interval. The initial conditions are automatically determined by d.c. analysis unless the *UIC* (use initial conditions) option is specified, and all sources that are not time dependent are set to their d.c. values. A request for transient analysis has the following general form:

```
.TRAN  tstep  tstop  <tstart  <tmax>>  <UIC>
```

Transient analysis always begins at time zero and continues to *tstop* with a printing or plotting time step of *tstep*. If the optional parameter *tstart* is included then the printing or plotting of results will be suppressed from time zero to *tstart*. Normally SPICE uses a computing time step that is the smaller of *tstep* and (*tstop* − *tstart*)/50. In some cases, however, this may be too large and the optional parameter *tmax* allows a time step that is smaller than *tstep* to be specified. The effect of including *UIC* is to start the transient analysis from a specified initial condition, rather than from a point obtained by d.c. analysis.

A request for transient analysis must be followed by a print or plot statement specifying the output variables. For example, the following statement would plot the voltages on nodes 1 and 2 over the time interval to 0 to 100 ms:

```
.TRAN  1MS  100MS
.PLOT  TRAN  V(1)  V(2)
```

In most cases one or more independent sources will be time dependent and SPICE provides five different source functions, of which the most important are sinusoidal and pulse. A sinusoidal source function has the general form:

```
SIN ( vo  va  freq  <td  <theta>> )
```

This generates a sinusoid of frequency *freq* and amplitude *va* with a d.c. offset of *vo*. A delay before the start of the sinusoid can be introduced by the optional parameter *td*; if *td* is omitted then there is no initial delay. An additional optional parameter *theta* causes the sinusoid to decay exponentially with time constant 1/*theta*. In most cases *theta* is omitted or set to zero to give a sinusoid of constant amplitude. Figure A.6 shows a waveform generated by a sinusoidal source function. A sinusoidal source function, together with an appropriate transient analysis request, can be used to determine the response of a network to a sinusoidal input. Consider, for example, the bridge rectifier shown in figure A.7.

Figure A.6
A sinusoidal
source function

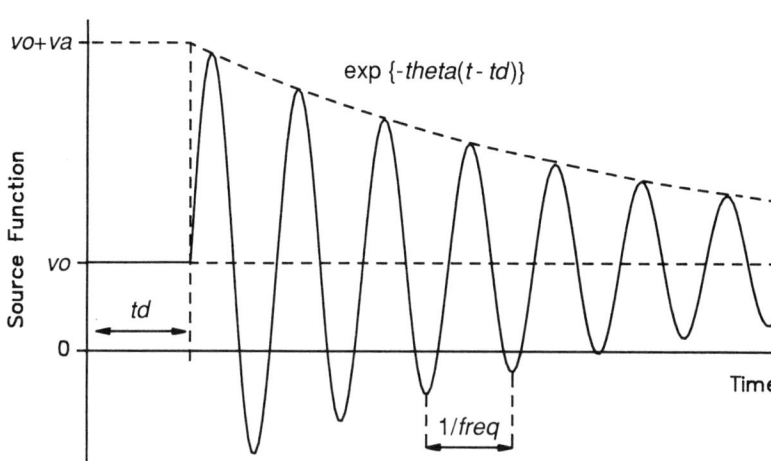

Figure A.7
A bridge
rectifier

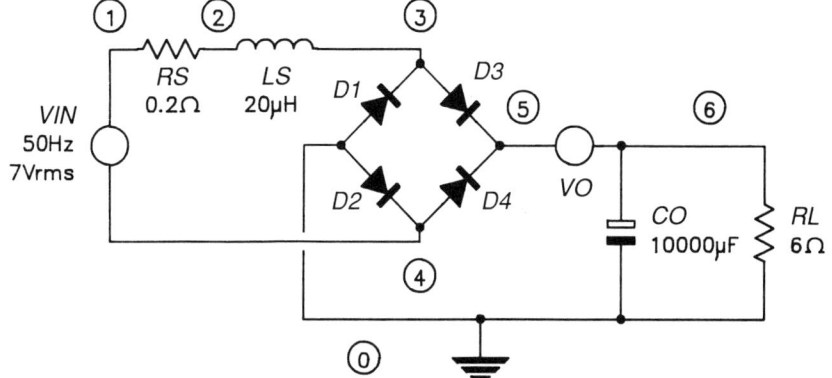

The resistance *RS*, inductance *LS* and the independent voltage source *VIN* represent the secondary winding of a transformer. A zero voltage source *VO* has been included to probe the current flowing in the rectifier diodes. In this example no delay is required and the sinusoid is of constant amplitude so that a suitable element description for *VIN* would be

```
VIN   1   4   SIN ( 0V   9.8V   50HZ )
```

A complete network description for the bridge rectifier, which plots the values of the diode current *I(VO)* and the load voltage *V6* over the first four cycles, is given below:

```
Bridge Rectifier
* Input:   7 Vrms 50 Hz sinusoid
* Output:  1 A nominal into 6 ohm load
.LIB mylib.cir
RS   1   2   0.2OHM
LS   2   3   20.0UH
D1   0   3   D1N4004
D2   0   4   D1N4004
D3   3   5   D1N4004
D4   4   5   D1N4004
CO   6   0   10000UF
RL   6   0   6.0OHM
VO   5   6   DC   0V
VIN  1   4   SIN ( 0.0V   9.8V   50HZ )
.TRAN   0.5MS   80.0MS   0.0MS   0.1MS
.PLOT   TRAN   V(6)   I(VO)
.END
```

Notice the use of the *tmax* parameter to force SPICE to use a computing time step of 0·1 ms, rather than a value equal to the plotting time step of 0·5 ms. Figure A.8 shows the results of this transient analysis.

Pulse source functions are used to generate either a single pulse or a train of identical pulses and have the general form:

```
PULSE ( v1  v2  <td  <tr  <tf  <pw  <per>>>>> )
```

Initial and pulsed values of the voltage or current pulse are specified by *v1* and *v2* respectively. A delay before the start of the pulse can be introduced by the optional parameter *td*; if *td* is omitted then there is no initial time delay. Pulse risetimes and falltimes are specified by *tr* and *tf*, which have default values of *tstep*. The pulse width is controlled by *pw* which has a default value of *tstop*; if *pw* is omitted then only the rising edge of the pulse

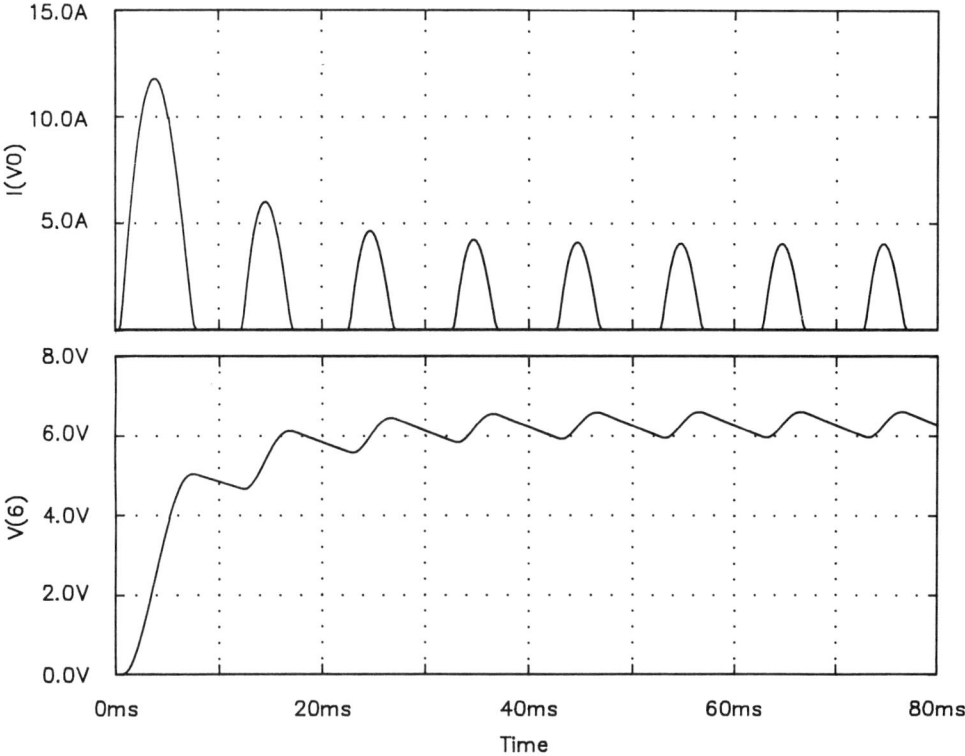

Figure A.8 Transient response of the bridge rectifier

is generated. Finally the pulse repetition period is specified by *per* which has a default value of *tstop*; if *per* is omitted then only a single pulse is generated. Figure A.9 shows the waveform produced by a pulse source function.

As an example of the use of pulse source functions, consider the network shown in figure A.10 which represents a type 7404 TTL inverter gate driving

Figure A.9
A pulse source
function

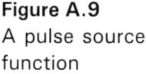
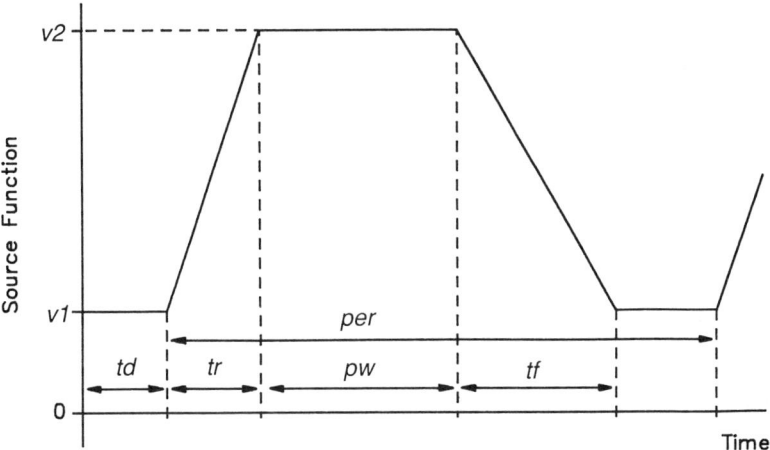

271

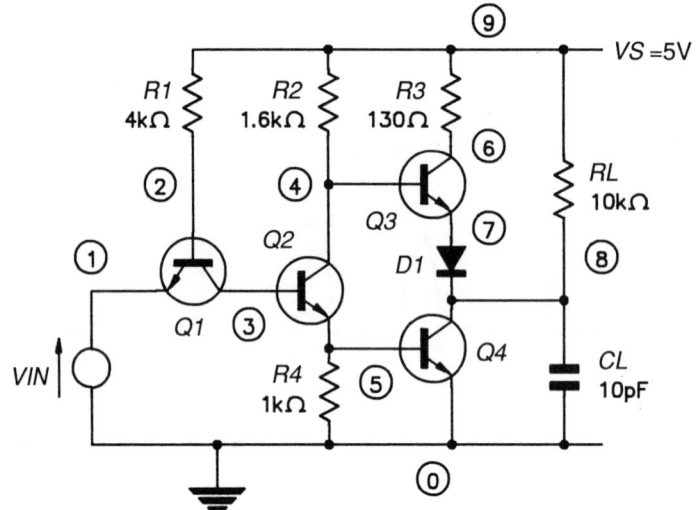

Figure A.10
A TTL inverter gate

a load consisting of a 10 kΩ resistance and a 10 pF capacitance. A single 50 ns wide logical 0 input pulse is to be used to test the response of the gate. The network description given below plots the input voltage *V(1)* and the output voltage *V(8)* over a time interval of 80 ns. Input voltage levels of 0 V and 4 V are used to represent logical 0 and logical 1 respectively.

```
   TTL Inverter 7404
* Transistor-Transistor Logic with Totem Pole Output
.MODEL  M1  NPN (BF=50  BR=.5  CJE=1PF  CJC=2PF  TF=0.1NS  TR=10NS)
.MODEL  M2  D (N=1.8  TT=1NS  CJO=1PF)
Q1   3   2   1   M1
Q2   4   3   5   M1
Q3   6   4   7   M1
Q4   8   5   0   M1
D1   7   8   M2
R1   2   9   4.0KOHM
R2   4   9   1.6KOHM
R3   5   0   1.0KOHM
R4   6   9   130OHM
RL   8   9   10KOHM
CL   8   0   10PF
VS   9   0   DC  5V
VIN  1   0   PULSE ( 4V  0V  5NS  5NS  10NS  45NS )
.TRAN  1.0NS  80NS  0.0NS  0.2NS
.PLOT  TRAN  V(1)  V(8)
.END
```

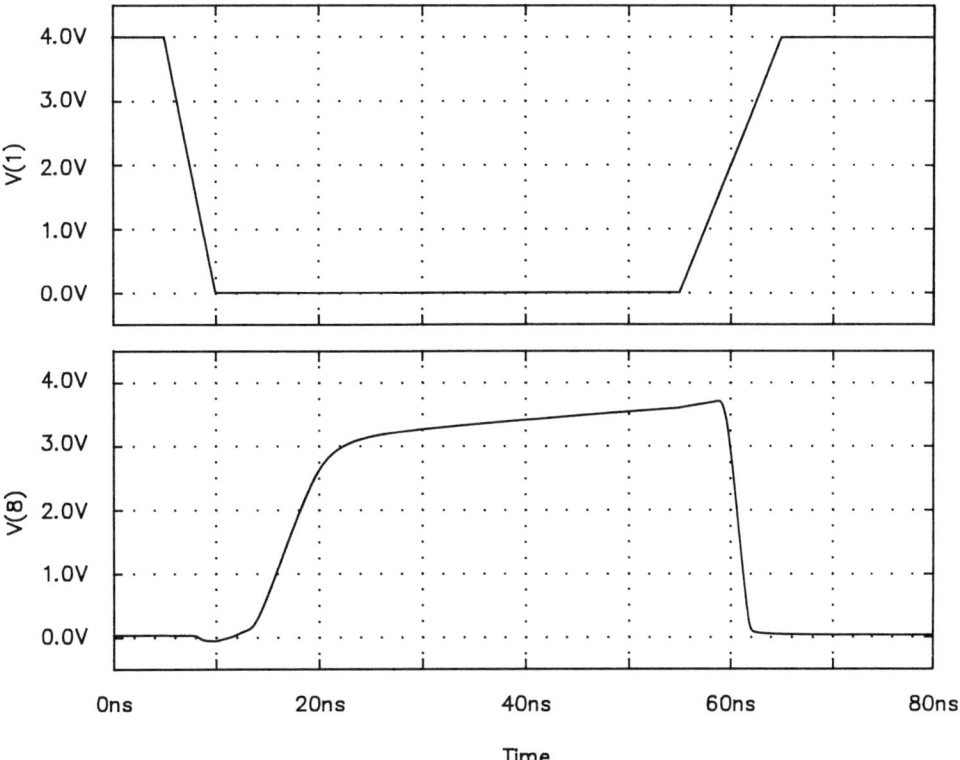

Figure A.11 Transient response of the inverter gate

As in the previous example a computing time step has been specified which is smaller than the plotting time step. Figure A.11 shows the results of this transient analysis.

SPICE will normally perform a d.c. analysis of the network prior to transient analysis. This can sometimes lead to difficulties, however. The algorithm used to obtain the d.c. solution may fail to converge, and when this happens SPICE prints the node voltages at the last iteration and terminates the analysis. Failure to converge in d.c. analysis can be the result of an error in specifying the network, but it can also occur in some correctly specified networks. In particular problems may be encountered with high-gain amplifiers and with networks having d.c. positive feedback.

It is possible to bypass the d.c. analysis and impose a set of initial conditions by the use of the *UIC* parameter in the transient analysis request. If *UIC* is active then SPICE uses the node voltages specified in an initial condition statement, or the branch voltages or currents specified by *IC*= parameters in element descriptions, as the starting point for transient analysis. Initial condition statements take precedence over *IC*= parameters and have the general form:

```
.IC   V(n1)=val1   V(n2)=val2   ...
```

273

where *n1* and *n2* are node numbers. For example, the initial condition statement given below sets node 1 to 4·0 V, node 2 to 1·5 V and node 3 to 2·5 V:

```
.IC  V(1)=4.0V  V(2)=1.5V  V(3)=2.5V
```

Another problem that can occur as a result of an initial d.c. analysis is metastability. This is seen in astable networks such as the emitter-coupled multivibrator shown in figure A.12. A suitable SPICE description of this network, together with a request for transient analysis over the time interval 0 to 200 μs, is given below:

```
Astable Multivibrator
* emitter-coupled multivibrator
.LIB mylib.cir
R1   1   2   4.7K
R2   2   0   4.7K
R3   4   0   39K
R4   5   0   82K
R5   3   1   18K
C1   4   5   1.0N
Q1   3   2   4   QBC182
Q2   1   3   5   QBC182
VS   1   0   DC   9.0V
.TRAN   1U   200U
.PLOT   TRAN   V(3)   V(4)   V(5)
.END
```

Figure A.12
An emitter-coupled multivibrator

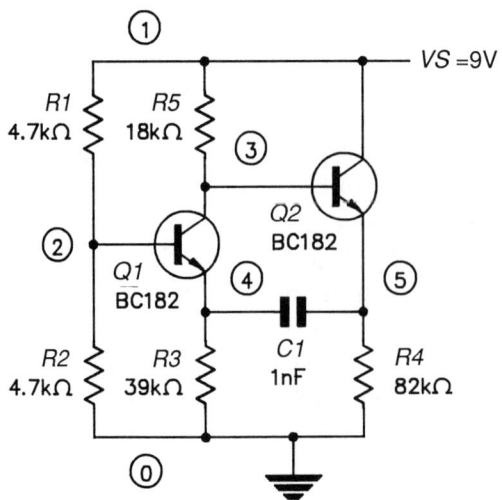

274

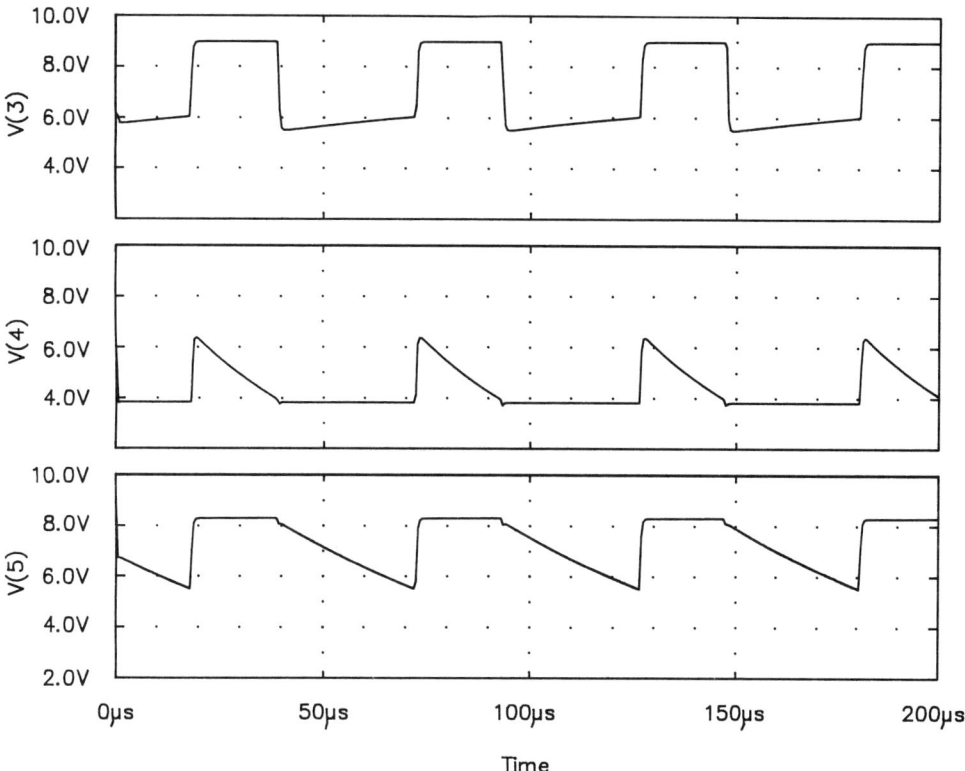

Figure A.13 Transient response of multivibrator

Surprisingly, when the transient analysis is performed it shows no signs of oscillation and the node voltages remain at their initial values. This is because at the operating point calculated by the initial d.c. analysis the network is metastable. Only a very small disturbance is required to initiate oscillation, and this will eventually be provided by rounding errors in the simulation; however the oscillation may take a considerable time to become established. This problem is overcome by specifying a non-metastable initial condition, for example:

```
.IC  V(4)=4.0V  V(5)=7.0V
.TRAN  1U  200U  UIC
```

With these modifications the oscillation starts up without delay, as illustrated in figure A.13.

A.10 Linear Dependent Sources

SPICE recognizes four different types of linear dependent source, namely voltage-controlled voltage sources (VCVS), voltage-controlled current sources (VCCS), current-controlled voltage sources (CCVS) and current-controlled current sources (CCCS). The general form of element descriptions for these linear dependent sources are respectively:

```
E*******   n1   n2   n3   n4   value
G*******   n1   n2   n3   n4   value
H*******   n1   n2   vnam   value
F*******   n1   n2   vnam   value
```

Output nodes are *n1* and *n2*; the sign convention is that for voltage sources *n1* is positive with respect to *n2*, and for current sources the current flows from *n1* through the source to *n2*. In the case of voltage-controlled sources the output is determined by the voltage difference between the positive controlling node *n3* and the negative controlling node *n4*. Current-controlled sources make use of an auxiliary voltage source *vnam* to sense the current at some point in the network and the output is proportional to this current. Normally *vnam* will be a zero voltage source and will have no direct effect on the operation of the network.

The ratio of output quantity to input quantity, which for a VCVS is the voltage gain, for a VCCS is the transconductance, for a CCVS is the transimpedance and for a CCCS is the current gain, is determined by the parameter *value*.

Voltage-controlled voltage sources can be used to represent operational amplifiers. A typical operational amplifier has a gain-bandwidth product of around 5 MHz. At an operating frequency of 1 kHz an operational amplifier with inputs nodes 1 and 2, and output node 3, could be represented by the VCVS element description:

```
E1   3   0   1   2   5000
```

This neglects the frequency variation of gain, the phase shift, and all non-linear effects such as output voltage, output current and slew-rate limiting.

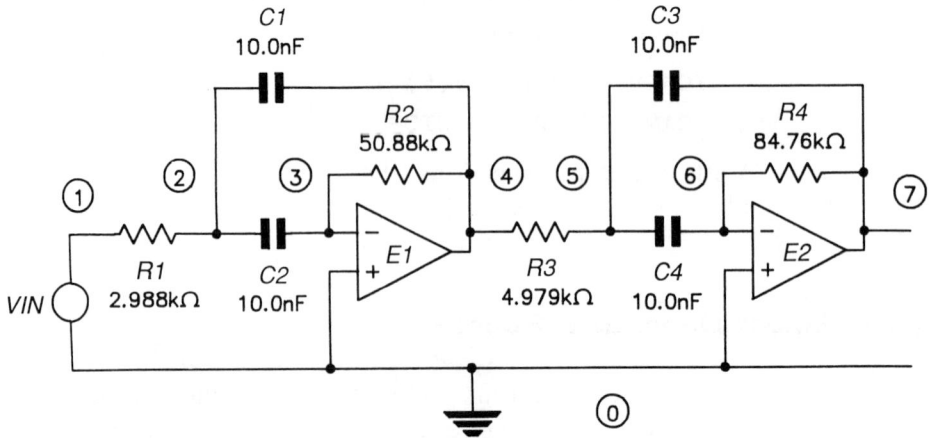

Figure A.14
A 4th-order
bandpass filter

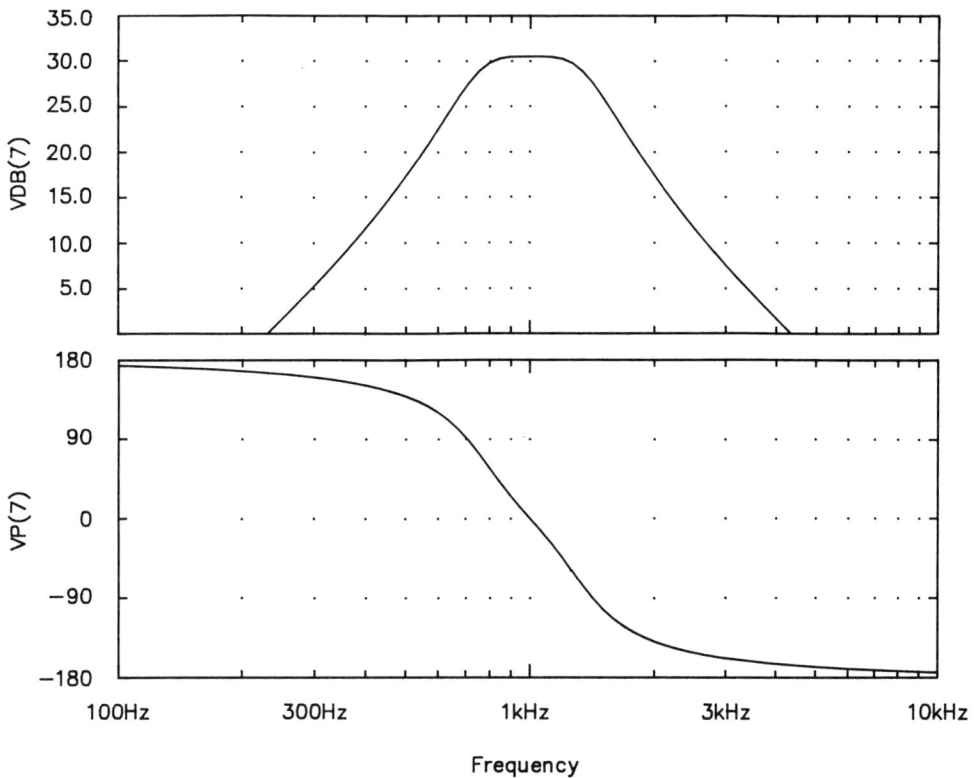

Figure A.15 Frequency response of the filter

Nevertheless it is often a useful approximation for analyzing linear networks containing operational amplifiers. Consider the 4th-order bandpass filter shown in figure A.14.

A network description of this filter, which plots the gain and the phase over the frequency range 100 Hz to 10 kHz, is given below:

```
Butterworth Filter
* 4th-order bandpass response
R1  1  2  2.988KOHM
R2  3  4  50.88KOHM
C1  2  4  10.0NF
C2  2  3  10.0NF
E1  4  0  0  3  5000
R3  4  5  4.979KOHM
R4  6  7  84.76KOHM
C3  5  7  10.0NF
C4  5  6  10.0NF
E2  7  0  0  6  5000
```

```
VIN   1   0   AC   1V
.AC   DEC   40   100   10000
.PLOT   AC   VDB(7)   VP(7)
.END
```

The result of performing this a.c. analysis is illustrated in figure A.15.

A.11 Sub-circuits

Networks often contain a number of identical sub-circuits, such as operational amplifiers or logic gates. For example a single-bit binary full adder might be constructed from nine 2-input NAND gates, each of which consisted of ten elements. It would be possible to write a network description for the binary adder in which each gate was defined individually in terms of its elements. However, this would involve a total of 90 elements. Fortunately SPICE provides a mechanism for defining sub-circuits, which can then be incorporated in the network description by single-line sub-circuit calls. A sub-circuit definition has the following general form:

```
.SUBCKT   subnam   n1   <n2   <n3   ..   >>
[definition]
.ENDS
```

The sub-circuit name is *subnam* and this is used in any subsequent sub-circuit calls. Nodes of the sub-circuit which are connected to other nodes in the main network are denoted by *n1*, *n2*, *n3*... and are known as external nodes. External nodes may not be the ground node, that is node 0. The sub-circuit definition can also include any number of internal nodes in addition to the external nodes. Internal nodes are strictly local to the sub-circuit definition and are not connected to the network outside, even if the node numbers are the same. There is one exception to this rule: if node 0 is used within the sub-circuit definition then this is taken to be the ground node of the main network.

A sub-circuit definition may include device models and other sub-circuit definitions. These are local and are not recognized outside the definition. Control statements, such as printing or plotting requests, are not allowed within sub-circuit definitions.

Sub-circuits are often used to represent operational amplifiers. It is possible to use a single VCVS for this purpose, as demonstrated in the previous section. A better representation, which gives the correct gain and phase characteristic, is shown in figure A.16.

The external nodes are 1, 2 and 3; nodes 1 and 2 are the operational amplifier inputs, and node 3 is the output. Node 4 is an internal node, and node 0 is the network ground. The input transconductance *GO* feeds the

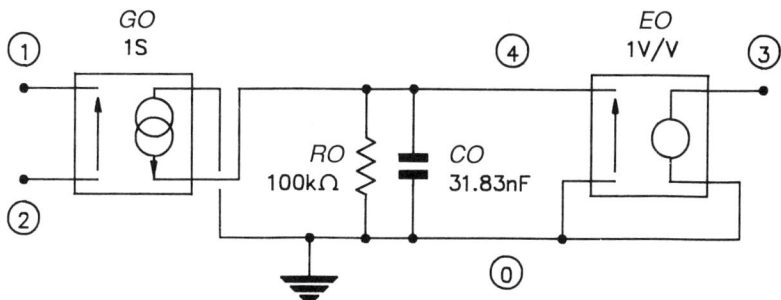

Figure A.16
Sub-circuit
representing an
operational
amplifier

parallel combination of *RO* and *CO*; output buffering is provided by the unity-gain voltage amplifier *EO*. Resistance *RO* determines the low-frequency gain A_0 of the amplifier:

$$A_0 = EO\ RO\ GO = 100000$$

and the gain-bandwidth product f_T depends on the capacitance *CO*;

$$f_T = \frac{EO\ GO}{2\pi\ CO} = 5\ \text{MHz}$$

A SPICE definition of the sub-circuit shown in figure A.16 is given below:

```
.SUBCKT   OPAMP   1   2   3
GO   0   4   1   2   1
RO   4   0   100KOHM
CO   4   0   31.83NF
EO   3   0   4   0   1
.ENDS  OPAMP
```

Sub-circuit calls have the following general form:

```
X*******   n1   <n2   <n3   ..   >>   subnam
```

Figure A.17
A precision
rectifier

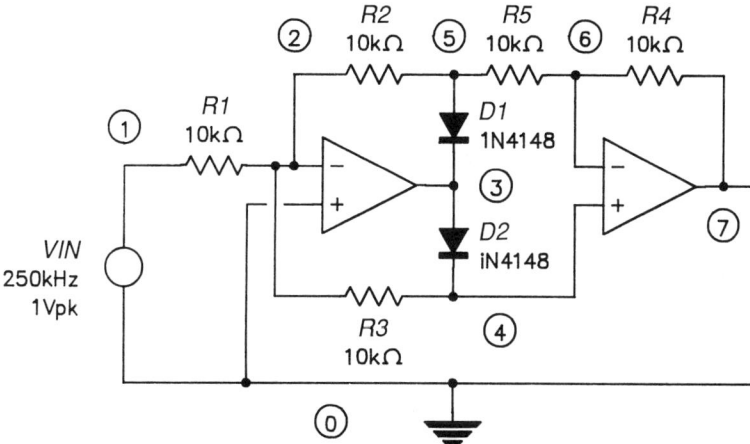

n1, n2, n3 ... are the nodes of the main network to which the external nodes of the sub-circuit are connected.

Figure A.17 shows a precision rectifier incorporating two operational amplifiers, each of gain–bandwidth product $f_T = 5$ MHz.

A SPICE network description of the precision rectifier, which performs a transient analysis over $2\frac{1}{2}$ cycles of the input sinusoid, is given below:

```
Precision Rectifier
* Precision rectifier using 2 opamps
.LIB mylib.cir
.SUBCKT  OPAMP  1  2  3
GO  0  4  1  2  1
RO  4  0  100KOHM
CO  4  0  31.83NF
EO  3  0  4  0  1
.ENDS OPAMP
R1  1  2  10KOHM
R2  2  5  10KOHM
R3  2  4  10KOHM
R4  6  7  10KOHM
R5  5  6  10KOHM
D1  5  3  D1N4148
D2  3  4  D1N4148
X1  0  2  3  OPAMP
X2  4  6  7  OPAMP
VIN  1  0  SIN ( 0.0V  1.0V  250KHZ )
.TRAN  0.05US  10.0US
.PLOT  TRAN  V(6)  V(7)
.END
```

The results of performing this transient analysis are illustrated in figure A.18, and the effects of the limited gain–bandwidth product of the operational amplifiers are clearly visible.

Although the simple representation of an operational amplifier shown in figure A.16 may be satisfactory for some purposes, it does not correctly model the non-linear behaviour. In particular, no account is taken of output slew-rate limiting, voltage limiting and current limiting. More sophisticated representations can be developed which better approximate the behaviour of real devices, but the increased complexity will result in slower SPICE simulations.

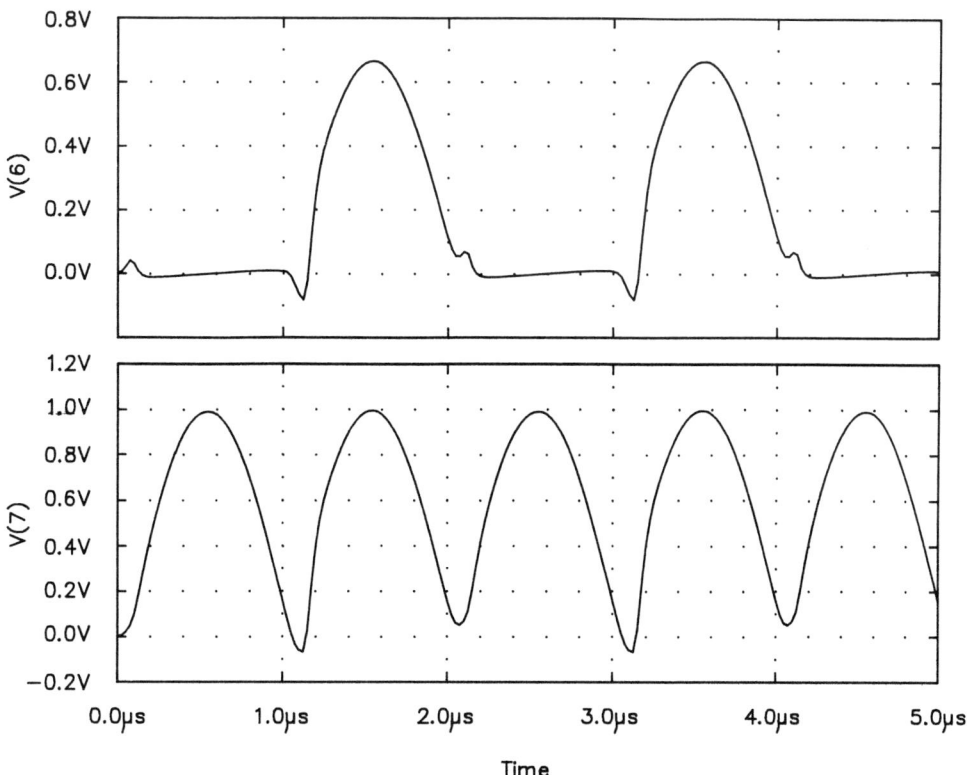

Figure A.18 Response of the precision rectifier

Appendix B: Modula-2 Definitions

The following constant and type definitions are used in the Modula-2 routines presented in this book.

```
CONST
    maxsize = 32;
    cmaxsize = 20;

    maxno = 99;
    maxbr = 80;

    infinity = 1.7E308;
    pi = 3.1415926535897932384;

TYPE
    function = PROCEDURE (REAL): REAL;

TYPE
    complex = RECORD re, im: REAL END;

    index = [0..maxsize];
    vector = ARRAY index OF REAL;
    matrix = ARRAY index OF vector;

    cindex = [0..cmaxsize];
    cvector = ARRAY cindex OF complex;
    cmatrix = ARRAY cindex OF cvector;
```

```
poly = ARRAY index OF REAL;
complexpoly = ARRAY index OF complex;
rational = RECORD a, b: poly;
                 order: index
             END;


node = [0..maxno];
branch = RECORD t1, t2: node END;
graph = ARRAY [1..maxbr] OF branch;


string = ARRAY [0 .. 8] OF CHAR;
ctype = (resistor, capacitor, inductor, conductor, vccs);

component = RECORD
     name: string;
     type: ctype;
     value: REAL
END;
circuit = ARRAY [1..maxbr] OF component;

statevar = RECORD A: matrix;
                  B, C: vector;
                  D: REAL;
                  order: index
              END;
```

Bibliography

Adby, P. R.: *Applied Circuit Theory*, Ellis Horwood (1980), ISBN 0-85312-071-4.

Adby, P. R. and Dempster M. A. H.: *Introduction to Optimization Methods*, Chapman and Hall (1974), ISBN 0-412-11040-7.

Banzaf, W.: *Computer-Aided Circuit Analysis Using SPICE*, Prentice-Hall International (1989), ISBN 0-13-168394-2.

Chen, W-K.: *Linear Networks and Systems*, Brooks/Cole Engineering Division (1983), ISBN 0-534-01343-0.

Fidler, J. K. and Nightingale, C.: *Computer Aided Circuit Design*, Thomas Nelson and Sons (1978), ISBN 0-17-761627-X.

King, K. N.: *Modula-2 A Complete Guide*, D. C. Heath and Company (1988), ISBN 0-669-11091-4.

Mastascusa, E. J.: *Computer-Assisted Network and System Analysis*, John Wiley & Sons (1988), ISBN 0-471-61231-6.

Spence, R. and Burgess, J. P.: *Circuit Analysis by Computer*, Prentice-Hall International (1986), ISBN 0-13-134024-7.

Spence, R. and Randeep, S. S.: *Tolerance Design of Electronic Circuits*, Addison Wesley Publishers (1988), ISBN 0-201-18242-4.

Wirth, N.: *Programming in Modula-2*, Springer Verlag (1985), ISBN 0-387-15078-1.

Index